Student Solution Manual for
Foundation Mathematics for the Physical Sciences
Student Solution Manual

This *Student Solution Manual* provides complete solutions to all the odd-numbered problems in *Foundation Mathematics for the Physical Sciences*. It takes students through each problem step by step, so they can clearly see how the solution is reached, and understand any mistakes in their own working. Students will learn by example how to arrive at the correct answer and improve their problem-solving skills.

K. F. RILEY read mathematics at the University of Cambridge and proceeded to a Ph.D. there in theoretical and experimental nuclear physics. He became a Research Associate in elementary particle physics at Brookhaven, and then, having taken up a lectureship at the Cavendish Laboratory, Cambridge, continued this research at the Rutherford Laboratory and Stanford; in particular he was involved in the experimental discovery of a number of the early baryonic resonances. As well as having been Senior Tutor at Clare College, where he has taught physics and mathematics for over 40 years, he has served on many committees concerned with the teaching and examining of these subjects at all levels of tertiary and undergraduate education. He is also one of the authors of *200 Puzzling Physics Problems* (Cambridge University Press, 2001).

M. P. HOBSON read natural sciences at the University of Cambridge, specialising in theoretical physics, and remained at the Cavendish Laboratory to complete a Ph.D. in the physics of star formation. As a Research Fellow at Trinity Hall, Cambridge, and subsequently an Advanced Fellow of the Particle Physics and Astronomy Research Council, he developed an interest in cosmology, and in particular in the study of fluctuations in the cosmic microwave background. He was involved in the first detection of these fluctuations using a ground-based interferometer. Currently a University Reader at the Cavendish Laboratory, his research interests include both theoretical and observational aspects of cosmology, and he is the principal author of *General Relativity: An Introduction for Physicists* (Cambridge University Press, 2006). He is also a Director of Studies in Natural Sciences at Trinity Hall and enjoys an active role in the teaching of undergraduate physics and mathematics.

Foundation Mathematics for the Physical Sciences

Student Solution Manual

K. F. RILEY
University of Cambridge

M. P. HOBSON
University of Cambridge

CAMBRIDGE
UNIVERSITY PRESS

CAMBRIDGE UNIVERSITY PRESS
Cambridge, New York, Melbourne, Madrid, Cape Town, Singapore,
São Paulo, Delhi, Dubai, Tokyo, Mexico City

Cambridge University Press
The Edinburgh Building, Cambridge CB2 8RU, UK

Published in the United States of America by Cambridge University Press, New York

www.cambridge.org
Information on this title: www.cambridge.org/9780521141048

First published 2011

Printed in the United Kingdom at the University Press, Cambridge

A catalogue record for this publication is available from the British Library

ISBN 978-0-521-14104-8 Paperback

Contents

Preface

For reasons that are explained in the preface to *Foundation Mathematics for the Physical Sciences* (*FMPS*), the text of the third edition of *Mathematical Methods for Physics and Engineering* (*MMPE*) (Cambridge: Cambridge University Press, 2006) by Riley, Hobson and Bence, after a number of additions and omissions, has been republished as two somewhat overlapping texts. *Essential Mathematical Methods for the Physical Sciences* contains most of the more advanced material, and specifically develops mathematical *methods* that can be applied throughout the physical sciences; *FMPS* is an augmented version of the more introductory material, principally concerned with mathematical *tools* rather than methods. The full text of *MMPE*, including all of the more specialised and advanced topics, is still available under its original title.

As in the third edition of *MMPE*, the penultimate subsection of each chapter of *FMPS* consists of a significant number of problems, nearly all of which are based on topics drawn from several sections of that chapter. Also as in the third edition, hints or outline answers are given in the final subsection, but only to the odd-numbered problems, leaving all even-numbered problems free to be set as unaided homework.

This book is the solutions manual for the problems in *FMPS*. For the two hundred and thirty plus *odd-numbered* problems it contains, complete solutions are available, to both students and their teachers, in the form of this manual; these are in addition to the hints or outline answers given in the main text. For each problem, the original question is reproduced and then followed by a fully worked solution. For those original problems that make internal reference to the main text or to other (even-numbered) problems not included in this solutions manual, the questions have been reworded, usually by including additional information, so that the questions can stand alone. Some further minor rewording has been included to improve the page layout.

In many cases the solution given is even fuller than one that might be expected of a good student who has understood the material. This is because we have aimed to make the solutions instructional as well as utilitarian. To this end, we have included comments that are intended to show how the plan for the solution is formulated and have provided the justifications for particular intermediate steps (something not always done, even by the best of students). We have also tried to write each individual substituted formula in the form that best indicates how it was obtained, before simplifying it at the next or a subsequent stage. Where several lines of algebraic manipulation or calculus are needed to obtain a final result, they are normally included in full; this should enable the student to determine whether an incorrect answer is due to a misunderstanding of principles or to a technical error.

As noted above, the original questions are reproduced in full, or in a suitably modified stand-alone form, at the start of each problem. Reference to the main text is not needed provided that standard formulae are known (and a set of tables is available for a few of the probability problems). This means that, although it is not its prime purpose, this manual could be used as a test or quiz book by a student who has learned, or thinks that they have learned, the material covered in the main text.

1 | Arithmetic and geometry

Powers and logarithms

1.1 Evaluate the following to 3 s.f.:

(a) e^π, (b) π^e, (c) $\log_{10}(\log_2 32)$, (d) $\log_2(\log_{10} 32)$.

Parts (a) and (b) do no more than test the understanding of notation, and are found directly using a calculator. (a) $e^\pi = 23.1$, and (b) $\pi^e = 22.5$. For the two other parts:

(c) $\log_{10}(\log_2 32) = \log_{10}(5) = 0.699$.
(d) $\log_2(\log_{10} 32) = \log_2(1.505)$. We therefore need the value of x that satisfies $2^x = 1.505$. To find it, take logarithms and obtain

$$x \ln 2 = \ln 1.505 \quad \Rightarrow \quad x = \frac{\ln 1.505}{\ln 2} = \frac{0.4088}{0.6931} = 0.590.$$

1.3 Find the number for which the cube of its square root is equal to twice the square of its cube root.

If a is the required number, then

$$a^{3/2} = 2a^{2/3} \quad \Rightarrow \quad 2 = a^{(3/2)-(2/3)} = a^{5/6}.$$

Now taking logarithms:

$$\tfrac{5}{6}\ln a = \ln 2 \quad \Rightarrow \quad a = e^{(6\ln 2)/5} = e^{0.83177\ldots} = 2.297\ldots$$

1.5 By applying the rationalisation procedure twice, show that

$$\frac{131}{3 - \sqrt{5} + \sqrt{7}} = 9 - 11\sqrt{5} + 7\sqrt{7} + 6\sqrt{35}.$$

Initially treating $\sqrt{7} - \sqrt{5}$ as one unit, we have

$$\frac{131}{3 - \sqrt{5} + \sqrt{7}} = \frac{131[3 - (\sqrt{7} - \sqrt{5})]}{9 - (\sqrt{7} - \sqrt{5})^2} = \frac{131[3 - (\sqrt{7} - \sqrt{5})]}{9 - 7 - 5 + 2\sqrt{35}}.$$

Since the denominator is now $-3 + 2\sqrt{35}$, as a second step we must multiply both numerator and denominator by $3 + 2\sqrt{35}$:

$$\frac{131}{3 - \sqrt{5} + \sqrt{7}} = \frac{131[3 - (\sqrt{7} - \sqrt{5})](3 + 2\sqrt{35})}{-9 + (4 \times 35)}$$

$$= \frac{131(9 - 3\sqrt{7} + 3\sqrt{5} + 6\sqrt{35} - 14\sqrt{5} + 10\sqrt{7})}{131}$$

$$= 9 - 11\sqrt{5} + 7\sqrt{7} + 6\sqrt{35}.$$

1.7 Solve the following for x:

(a) $x = 1 + \ln x$,   (b) $\ln x = 2 + 4 \ln 3$,   (c) $\ln(\ln x) = 1$.

(a) By inspection of either the original equation or its exponentiated form, $e^x = e^1 e^{\ln x} = ex$, we conclude that $x = 1$.

(b) By exponentiation, $x = e^{2 + 4 \ln 3} = e^2 e^{4 \ln 3} = e^2 3^4 = 81 e^2 = 598.5$.

(c) $\ln(\ln x) = 1 \quad \Rightarrow \quad \ln x = e \quad \Rightarrow \quad x = e^e = 15.15$.

1.9 Express $(2n + 1)(2n + 3)(2n + 5) \ldots (4n - 3)(4n - 1)$ in terms of factorials.

Denoting the expression by $f(n)$,

$$f(n) = \frac{(4n)!}{(2n)!} \frac{1}{(2n + 2)(2n + 4) \ldots (4n - 2)(4n)}$$

$$= \frac{(4n)!}{(2n)! \, (n + 1)(n + 2) \ldots (2n - 1)(2n) \, 2^n}$$

$$= \frac{(4n)! \, n!}{(2n)! \, (2n)! \, 2^n}.$$

1.11 Measured quantities x and y are known to be connected by the formula

$$y = \frac{ax}{x^2 + b},$$

where a and b are constants. Pairs of values obtained experimentally are

x:	2.0	3.0	4.0	5.0	6.0
y:	0.32	0.29	0.25	0.21	0.18

Use these data to make best estimates of the values of y that would be obtained for (a) $x = 7.0$, and (b) $x = -3.5$. As measured by fractional error, which estimate is likely to be the more accurate?

In order to use this limited data to best advantage when estimating a and b graphically, the equation needs to be arranged in the linear form $v = mu + c$, since a straight-line graph is much the easiest form from which to extract parameters. The given equation can be arranged as

$$\frac{x}{y} = \frac{x^2}{a} + \frac{b}{a},$$

which is represented by a line with slope a^{-1} and intercept b/a when x^2 is used as the independent variable and x/y as the dependent one. The required tabulation is:

x	2.0	3.0	4.0	5.0	6.0
y	0.32	0.29	0.25	0.21	0.18
x^2	4.0	9.0	16.0	25.0	36.0
x/y	6.25	10.34	16.00	23.81	33.33

Plotting these data as a graph for $0 \le x^2 \le 40$ produces a straight line (within normal plotting accuracy). The line has a slope

$$\frac{1}{a} = \frac{28.1 - 2.7}{30.0 - 0.0} = 0.847 \quad \Rightarrow \quad a = 1.18.$$

The intercept is at $x/y = 2.7$, and, as this is equal to b/a, it follows that $b = 2.7 \times 1.18 = 3.2$. In fractional terms this is not likely to be very accurate, as $b \ll x^2$ for all but two of the x-values used.

    (a) For $x = 7.0$, the estimated value of y is

$$y = \frac{1.18 \times 7.0}{49.0 + 3.2} = 0.158.$$

    (b) For $x = -3.5$, the estimated value of y is

$$y = \frac{1.18 \times (-3.5)}{12.25 + 3.2} = -0.267.$$

Although as a graphical extrapolation estimate (b) is further removed from the measured values, it is likely to be the more accurate because, using the fact that $y(-x) = -y(x)$, it is effectively obtained by (visual) *inter*polation amongst measured data rather than by *extra*polation from it.

1.13 The variation with the absolute temperature T of the thermionic emission current i from a heated surface (in the absence of space charge effects) is said to be given by

$$i = AT^2 e^{-BT},$$

where A and B are both independent of T. How would you plot experimental measurements of i as a function of T so as to check this relationship and then extract values for A and B?

The equation can be rearranged to read $\ln(i/T^2) = \ln A - BT$, and so $y = \ln(i/T^2)$ should be plotted against T, obtaining a straight-line graph if the relationship is valid. If so, the (negative) slope of the graph gives B and the intercept y_0 on the y-axis gives A as $A = e^{y_0}$.

Dimensions

1.15 Three very different lengths that appear in quantum physics and cosmology are the Planck length ℓ_p, the Compton wavelength λ_m, and the Schwarzchild radius r_s. Given that

$$\ell_p = \sqrt{\frac{hG}{2\pi c^n}}, \qquad \lambda_m = \frac{h}{mc}, \qquad r_s = \frac{2GM}{c^2},$$

where m and M are masses, calculate the dimensions of the gravitational constant G and those of the Planck constant h. Deduce the value of n in the formula for the Planck length.

Remembering that numerical constants do not contribute to 'dimensional equations', from the Schwarzschild radius formula we have

$$[G] = \left[\frac{c^2 r_s}{M}\right] = \frac{(LT^{-1})^2 L}{M} = L^3 T^{-2} M^{-1}.$$

From the Compton wavelength formula it follows that, $[h] = [\lambda mc] = L^2 T^{-1} M$. Finally, from the Planck length formula

$$[c^n] = \left[\frac{hG}{l_p^2}\right] = \frac{L^2 T^{-1} M \, L^3 T^{-2} M^{-1}}{L^2} = L^3 T^{-3},$$

from which, since $[c] = LT^{-1}$, it follows that $n = 3$.

1.17 According to Bohr's theory of the hydrogen atom, the ionisation energy of hydrogen is $m_e e^4/8\epsilon_0^2 h^2$. Using Appendix A, show that this expression does have the dimensions of an energy and that its value when expressed in electron-volts is 13.8 eV.

The dimensions of ϵ_0 seem difficult to determine, but one common equation containing it is that for the force F between two charges separated by a distance r, i.e.

$$F = \frac{q_1 q_2}{4\pi \epsilon_0 r^2} \qquad \Rightarrow \qquad [F] = \frac{[q]^2}{[\epsilon_0] L^2} \qquad (*).$$

The dimensions of the given expression for the ionization energy E contain $[q]^4/[\epsilon_0]^2$, and by squaring $(*)$ we deduce that they are $([F]L^2)^2$. Using this, and the result from Problem 1.15 that $[h] = ML^2 T^{-1}$, the dimensions of $[E]$ are given by

$$[E] = \frac{M}{(ML^2 T^{-1})^2} \times (MLT^{-2} \times L^2)^2 = \frac{M^3 L^6 T^{-4}}{L^4 T^{-2} M^2} = \frac{ML^2}{T^2} = [\text{energy}].$$

Substituting the numerical values given in Appendix A yields a value of 2.2×10^{-18} J, which is $2.2 \times 10^{-18} \div 1.60 \times 10^{-19} = 13.8$ when expressed in electron-volts.

1.19 The electrical conductivity σ of a metal is measured in siemens per metre (S m^{-1}), where 1 S is the unit of conductance of an electrical component and is equivalent to 1 A V^{-1}. The Wiedemann–Franz law states that at absolute temperature T, and under certain conditions, σ is related to the thermal conductivity λ of the metal by the equation

$$\frac{\lambda}{\sigma T} = \frac{\pi^2}{3} \left(\frac{k}{e}\right)^2.$$

Verify that this equation is dimensionally acceptable and, using Appendix A, estimate the thermal conductivity of copper at room temperature, given that its electrical conductivity is 5.6×10^7 S m^{-1}.

Dealing first with the LHS of the equation, the thermal conductivity is measured in joules per square metre per second for a unit temperature gradient (or watts per metre per kelvin). Its dimensions are therefore given by

$$[\lambda] = \frac{ML^2T^{-2}\,L^{-2}\,T^{-1}}{\theta L^{-1}} = MLT^{-3}\theta^{-1}.$$

From the formula watts $=$ volts $\times$ amps, we conclude that the dimensions of voltage are $ML^2T^{-2}T^{-1} \times I^{-1}$, and that those of a siemens are therefore $I^2T^3M^{-1}L^{-2}$. This gives the dimensions of σ as $[\sigma] = M^{-1}L^{-3}T^3I^2$, from which it follows that $[\lambda/\sigma T] = M^2L^4T^{-6}I^{-2}\theta^{-2}$; recall that T in the given equation is a temperature, not a time.

On the RHS, since the Boltzmann constant k has units of joules per kelvin, and charge $=$ current $\times$ time, $[k/e] = (ML^2T^{-2}\theta^{-1})/(IT)$ leading to $[(k/e)^2] = M^2L^4T^{-6}I^{-2}\theta^{-2}$. Comparison with the result for the LHS shows that the stated formula is dimensionally acceptable. Taking room temperature as 293 K, and substituting other values from Appendix A, gives an estimate for λ of 402 W m^{-1} K^{-1}.

1.21 The following is a student's proposed formula for the energy flux S (the magnitude of the so-called Poynting vector) associated with an electromagnetic wave in a vacuum, the electric field strength of the wave being E and the associated magnetic flux density being B:

$$S = \frac{1}{2}\left[\left(\frac{\epsilon_0}{\mu_0}\right)^{1/2} E^2 + \left(\frac{\mu_0}{\epsilon_0}\right)^{1/2} B^2\right].$$

The dimensions of ϵ_0, the permittivity of free space, are $M^{-1}L^{-3}T^4I^2$, and those of its permeability μ_0 are $MLT^{-2}I^{-2}$. Given, further, that the force acting on a rod of length ℓ that carries a current I at right angles to a field of magnetic flux density B is $BI\ell$, determine whether the student's formula could be correct and, if not, locate the error as closely as possible.

From the force on a current-carrying rod,

$$[B] = \frac{[\text{force}]}{[\text{current}] \times L} = \frac{MLT^{-2}}{IL} = MT^{-2}I^{-1},$$

whilst energy flux, measured in joules per square metre per second, has dimensions $(ML^2T^{-2})L^{-2}T^{-1} = MT^{-3}$.

Since the electric field E has dimensions $[V]L^{-1}$, and from the solution to Problem 1.19, $[V] = ML^2T^{-3}I^{-1}$, the dimensions of E are $MLT^{-3}I^{-1}$. Those of the 'E^2' term are therefore

$$\left(\frac{M^{-1}L^{-3}T^4I^2}{MLT^{-2}I^{-2}}\right)^{1/2}(MLT^{-3}I^{-1})^2 = MT^{-3}.$$

The corresponding calculation for the 'B^2' term is

$$\left(\frac{MLT^{-2}I^{-2}}{M^{-1}L^{-3}T^4I^2}\right)^{1/2}(MT^{-2}I^{-1})^2 = L^2M^3T^{-7}I^{-4}.$$

Thus the electric term is compatible with an energy flux, but the magnetic one is not, and the error is almost certainly in the latter.

Binomial expansion

1.23 Evaluate those of the following that are defined: (a) $^5C_3$, (b) $^3C_5$, (c) $^{-5}C_3$, (d) $^{-3}C_5$.

(a) $^5C_3 = \dfrac{5!}{3!\,2!} = 10.$

(b) $^3C_5$. This is not defined as $5 > 3 > 0$.

For (c) and (d) we will need to use the identity

$$^{-m}C_k = (-1)^k\frac{m(m+1)\cdots(m+k-1)}{k!} = (-1)^k\,{}^{m+k-1}C_k.$$

(c) $^{-5}C_3 = (-1)^3\,{}^{5+3-1}C_3 = -\dfrac{7!}{3!\,4!} = -35.$

(d) $^{-3}C_5 = (-1)^5\,{}^{5+3-1}C_5 = -\dfrac{7!}{5!\,2!} = -21.$

1.25 By applying the binomial expansion directly to the identity

$$(x+y)^p(x+y)^q \equiv (x+y)^{p+q},$$

prove the result

$$\sum_{t=0}^{r}{}^pC_{r-t}\,{}^qC_t = {}^{p+q}C_r = \sum_{t=0}^{r}{}^pC_t\,{}^qC_{r-t}$$

which gives a formula for combining terms from two sets of binomial coefficients in a particular way (a kind of 'convolution', for readers who are already familiar with this term).

First, we write each term in the form

$$(x + y)^m = \sum_{s=0}^{m} {}^m C_s x^s y^{m-s},$$

where m represents, in turn, p, q and $p + q$.

Next we consider all the terms in the product of sums on the LHS that lead to terms containing x^r. If the first sum contributes a term containing x^{r-t}, with $0 \le t \le r$, then the second sum must contribute one containing x^t. The power of y that is in the same product term will be $y^{p-(r-t)} \times y^{q-t} = y^{p+q-r}$. The full form of the term, including the relevant binomial coefficients, is therefore ${}^p C_{r-t} x^{r-t} y^{p-r+t} \times {}^q C_t x^t y^{q-t}$.

The sum of all these terms over $t = 0, 1, \ldots, r$ must also give the coefficient of x^r in the expansion of $(x + y)^{p+q}$, i.e. ${}^{p+q} C_r$; this establishes the left-hand equality. The right-hand equality follows, either by symmetry or by interchanging the roles of p and q.

Trigonometric identities

1.27 Prove that

$$\cos \frac{\pi}{12} = \frac{\sqrt{3} + 1}{2\sqrt{2}}$$

by considering

(a) the sum of the sines of $\pi/3$ and $\pi/6$,
(b) the sine of the sum of $\pi/3$ and $\pi/4$.

(a) Using

$$\sin A + \sin B = 2 \sin \left(\frac{A + B}{2} \right) \cos \left(\frac{A - B}{2} \right),$$

we have

$$\sin \frac{\pi}{3} + \sin \frac{\pi}{6} = 2 \sin \frac{\pi}{4} \cos \frac{\pi}{12},$$

$$\frac{\sqrt{3}}{2} + \frac{1}{2} = 2 \frac{1}{\sqrt{2}} \cos \frac{\pi}{12},$$

$$\cos \frac{\pi}{12} = \frac{\sqrt{3} + 1}{2\sqrt{2}}.$$

(b) Using, successively, the identities

$$\sin(A + B) = \sin A \cos B + \cos A \sin B,$$

$$\sin(\pi - \theta) = \sin \theta,$$

$$\cos(\tfrac{1}{2}\pi - \theta) = \sin \theta,$$

we obtain

$$\sin\left(\frac{\pi}{3} + \frac{\pi}{4}\right) = \sin\frac{\pi}{3}\cos\frac{\pi}{4} + \cos\frac{\pi}{3}\sin\frac{\pi}{4},$$

$$\sin\frac{7\pi}{12} = \frac{\sqrt{3}}{2}\frac{1}{\sqrt{2}} + \frac{1}{2}\frac{1}{\sqrt{2}},$$

$$\sin\frac{5\pi}{12} = \frac{\sqrt{3}+1}{2\sqrt{2}},$$

$$\cos\frac{\pi}{12} = \frac{\sqrt{3}+1}{2\sqrt{2}}.$$

As they must, the two methods give the same answer.

1.29 Find the real solutions of

(a) $3\sin\theta - 4\cos\theta = 2$,
(b) $4\sin\theta + 3\cos\theta = 6$,
(c) $12\sin\theta - 5\cos\theta = -6$.

We use the result that if

$$a\sin\theta + b\cos\theta = k$$

then

$$\theta = \sin^{-1}\left(\frac{k}{K}\right) - \phi,$$

where

$$K^2 = a^2 + b^2 \quad \text{and} \quad \phi = \tan^{-1}\frac{b}{a}.$$

Recalling that the inverse sine yields two values and that the individual signs of a and b have to be taken into account, we have

(a) $k = 2$, $K = \sqrt{3^2 + 4^2} = 5$, $\phi = \tan^{-1}(-4/3)$ and so

$$\theta = \sin^{-1}\frac{2}{5} - \tan^{-1}\frac{-4}{3} = 1.339 \text{ or } -2.626.$$

(b) $k = 6$, $K = \sqrt{4^2 + 3^2} = 5$. Since $k > K$, there is no solution for a real angle θ.
(c) $k = -6$, $K = \sqrt{12^2 + 5^2} = 13$, $\phi = \tan^{-1}(-5/12)$ and so

$$\theta = \sin^{-1}\frac{-6}{13} - \tan^{-1}\frac{-5}{12} = -0.0849 \text{ or } -2.267.$$

1.31 Find all the solutions of

$$\sin\theta + \sin 4\theta = \sin 2\theta + \sin 3\theta$$

that lie in the range $-\pi < \theta \le \pi$. What is the multiplicity of the solution $\theta = 0$?

Using

$$\sin A + \sin B = 2 \sin \left(\frac{A+B}{2} \right) \cos \left(\frac{A-B}{2} \right)$$

$$\text{and} \quad \cos A - \cos B = -2 \sin \left(\frac{A+B}{2} \right) \sin \left(\frac{A-B}{2} \right),$$

and recalling that $\cos(-\phi) = \cos(\phi)$, the equation can be written successively as

$$2 \sin \frac{5\theta}{2} \cos \left(-\frac{3\theta}{2} \right) = 2 \sin \frac{5\theta}{2} \cos \left(-\frac{\theta}{2} \right),$$

$$\sin \frac{5\theta}{2} \left(\cos \frac{3\theta}{2} - \cos \frac{\theta}{2} \right) = 0,$$

$$-\sin \frac{5\theta}{2} \sin \theta \sin \frac{\theta}{2} = 0.$$

The first factor gives solutions for θ of $-4\pi/5$, $-2\pi/5$, 0, $2\pi/5$ and $4\pi/5$. The second factor gives rise to solutions 0 and π, whilst the only value making the third factor zero is $\theta = 0$. The solution $\theta = 0$ appears in each of the sets and so has multiplicity 3.

Inequalities

1.33 Starting from the double inequality $n - 1 < n < n + 1$, show that, for $n \geq 1$,

$$\sqrt{n} - \sqrt{n-1} > \frac{1}{2\sqrt{n}} > \sqrt{n+1} - \sqrt{n}.$$

Deduce that $\sum_{n=1}^{99} n^{-1/2}$ lies in the interval $(18, 6\sqrt{11})$.

As all the terms in the original inequality are non-negative, we can take the square roots of each term and maintain the double inequality. Doing this, and adding $\sqrt{n}$ to each term, we obtain

$$\sqrt{n-1} + \sqrt{n} < 2\sqrt{n} < \sqrt{n+1} + \sqrt{n}.$$

In view of the central term in the required result we now take reciprocals (which involves reversing the inequalities):

$$\frac{1}{\sqrt{n-1} + \sqrt{n}} > \frac{1}{2\sqrt{n}} > \frac{1}{\sqrt{n+1} + \sqrt{n}}.$$

Finally, we rationalise the two outer expressions using the identity

$$\frac{1}{a+b} = \frac{a-b}{(a+b)(a-b)} = \frac{a-b}{a^2 - b^2}$$

and obtain

$$\frac{\sqrt{n-1}-\sqrt{n}}{(n-1)-n} > \frac{1}{2\sqrt{n}} > \frac{\sqrt{n+1}-\sqrt{n}}{(n+1)-n},$$

$$\sqrt{n}-\sqrt{n-1} > \frac{1}{2\sqrt{n}} > \sqrt{n+1}-\sqrt{n}.$$

We can now sum each term of the double inequality from $n=1$ to $n=99$ without invalidating it:

$$\sum_{n=1}^{99}(\sqrt{n}-\sqrt{n-1}) > \sum_{n=1}^{99}\left(\frac{1}{2\sqrt{n}}\right) > \sum_{n=1}^{99}(\sqrt{n+1}-\sqrt{n}),$$

$$\sqrt{99}-0 > \sum_{n=1}^{99}\left(\frac{1}{2\sqrt{n}}\right) > \sqrt{100}-\sqrt{1},$$

$$3\sqrt{11} > \sum_{n=1}^{99}\left(\frac{1}{2\sqrt{n}}\right) > 9,$$

$$6\sqrt{11} > \sum_{n=1}^{99}\left(\frac{1}{\sqrt{n}}\right) > 18.$$

1.35 By finding suitable values for A and B in the function $A(x-3)^2 + B(x-7)^2$ show that $f(x) = 6x^2 - 68x + 214$ cannot be zero for any value of x. Further, by rearranging the expression for $f(x)$, show that its actual minimum value is $64/3$.

Equating the coefficients of x^2 and x in $6x^2 - 68x + 214 = A(x-3)^2 + B(x-7)^2$ gives

$$A + B = 6,$$

$$-6A - 14B = -68,$$

yielding $A = 2$ and $B = 4$. Thus, $f(x) = 2(x-3)^2 + 4(x-7)^2$; this is necessarily > 0, since each term is ≥ 0 and they cannot be zero together.
Write

$$f(x) = 6\left[x^2 - \frac{68}{6}x + \left(\frac{34}{6}\right)^2\right] - 6\left(\frac{34}{6}\right)^2 + 214.$$

The first term is a perfect square, which is therefore ≥ 0, and the second term gives the minimum value as $214 - 6(34/6)^2 = 64/3$.

1.37 For the pair of inequalities

$$ax + by > e,$$

$$cx + dy < f,$$

in which $a, b, \ldots, f$ are all positive, consider the following calculation:

$$d(ax + by) > de, \quad b(cx + dy) < bf, \quad \text{using (1e) and (2e)}$$

$$\Rightarrow \quad d(ax + by) - b(cx + dy) > de - bf, \quad \text{using (1d)}$$

$$\Rightarrow \quad x(ad - bc) > de - bf,$$

$$\Rightarrow \quad x > \frac{de - bf}{ad - bc}. \quad (*)$$

For the two particular cases

$$\text{(i)} \begin{cases} 2x + 3y > 12, \\ 3x + 4y < 25 \end{cases} \quad \text{and} \quad \text{(ii)} \begin{cases} 5x + 4y > 29 \\ 3x + 4y < 25 \end{cases}$$

verify that for $x = 5$ and $y = 2$ all four inequalities are valid. Now show that deduction $(*)$ is not a valid statement in case (i), although in case (ii) it is. Explain why this is so and how the calculation should be corrected in the former case.

For $x = 5$ and $y = 2$,

$$\text{(i)} \begin{cases} 10 + 6 > 12 \\ 15 + 8 < 25 \end{cases} \quad \text{and} \quad \text{(ii)} \begin{cases} 25 + 8 > 29 \\ 15 + 8 < 25 \end{cases}$$

and so all four inequalities are satisfied.

Now substituting in the 'result' gives

$$\text{(i)} \ 5 > \frac{(4)(12) - (3)(25)}{(2)(4) - (3)(3)} = 27 \quad \text{(ii)} \ 5 > \frac{(4)(29) - (4)(25)}{(5)(4) - (3)(4)} = 2.$$

Whilst result (ii) is valid, result (i) is clearly not.

In case (i), the expression '$ad - bc$' has the value $(2)(4) - (3)(3) = -1$ and is therefore negative; this means that the inequality should have been reversed when the line $(*)$ was derived. In case (ii), the corresponding factor is $(5)(4) - (4)(3) = 8$, i.e. positive, and the division by $ad - bc$ was carried out correctly.

1.39 Determine the range(s) of x that simultaneously satisfy the three inequalities

(i) $x^2 - 6 \leq x$,   (ii) $|x - 1| \geq 1$,   (iii) $x^2 + 2 > 3$.

We must find the range, or ranges, of x allowed by each of the inequalities, and then determine whether there is any range common to all three:

(i) This can be rewritten as $x^2 - x - 6 \leq 0$, i.e. $(x - 3)(x + 2) \leq 0$. This implies that $-2 \leq x \leq 3$.

(ii) The second inequality implies that either $x \leq 0$ or $x \geq 2$.

(iii) The final inequality requires that either $x < -1$ or $x > 1$.

Taken together the three requirements are satisfied only if $-2 \leq x < -1$ or $2 \leq x \leq 3$.

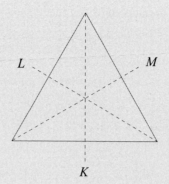

Figure 1.1 Reflections in the three perpendicular bisectors of the sides of an equilateral triangle take the triangle into itself.

Commutativity and associativity

1.41 The 'group' of symmetry operations on an equilateral triangle consists of (clockwise) rotations (about an axis perpendicular to its plane and passing through its centre) by 0, $2\pi/3$ and $-2\pi/3$, and denoted respectively by A, B and C, together with the reflections of the same triangle in the bisectors of each of the three sides, denoted by K, L and M (see Figure 1.1).

The product $X \odot Y$ is defined as the single element from amongst $A, B, \ldots, M$ that is equivalent to first applying operation Y to the triangle, and then applying operation X to the result. Thus, as examples, $A \odot X = X = X \odot A$ for any A, $B \odot C = A$, $B \odot K = M$ and $L \odot C = K$. These results have been entered into the 6×6 'multiplication table' for the group, which has row-headings X and column-headings Y:

$x = $ \ $y = $	A	B	C	K	L	M
A	A	B	C	K	L	M
B	B		A	M		
C	C					
K	K				K	
L	L					
M	M					

Complete the table and then use it to decide whether $\odot$ is (a) commutative, and (b) associative.

(a) The completed table is

$x = $ \ $y = $	A	B	C	K	L	M
A	A	B	C	K	L	M
B	B	C	A	M	K	L
C	C	A	B	L	M	K
K	K	L	M	A	B	C
L	L	M	K	C	A	B
M	M	K	L	B	C	A

Operation $\odot$ is clearly not commutative. For example $K \odot L = B$ but $L \odot K = C$.

(b) For product $X \odot Y \odot Z$, treated as both $X \odot (Y \odot Z)$ and $(X \odot Y) \odot Z$, consider a sample of each of the following cases:

(i) One or more elements is A;

(ii) $B \odot C \odot P$ and $B \odot P \odot C$, where P is one of K, L, and M;

(iii) $K \odot L \odot Q$ and $K \odot Q \odot L$, where Q is either B or C.

In all cases it is found that $X \odot (Y \odot Z) = (X \odot) Y \odot Z$, showing that the operation is associative.

2 Preliminary algebra

Polynomial equations

2.1 It can be shown that the polynomial

$$g(x) = 4x^3 + 3x^2 - 6x - 1$$

has turning points at $x = -1$ and $x = \frac{1}{2}$ and three real roots altogether. Continue an investigation of its properties as follows.

(a) Make a table of values of $g(x)$ for integer values of x between -2 and 2. Use it and the information given above to draw a graph and so determine the roots of $g(x) = 0$ as accurately as possible.

(b) Find one accurate root of $g(x) = 0$ by inspection and hence determine precise values for the other two roots.

(c) Show that $f(x) = 4x^3 + 3x^2 - 6x - k = 0$ has only one real root unless $-5 \leq k \leq \frac{7}{4}$.

(a) Straightforward evaluation of $g(x)$ at integer values of x gives the following table:

x	-2	-1	0	1	2
$g(x)$	-9	4	-1	0	31

(b) It is apparent from the table alone that $x = 1$ is an exact root of $g(x) = 0$ and so $g(x)$ can be factorised as $g(x) = (x - 1)h(x) = (x - 1)(b_2x^2 + b_1x + b_0)$. Equating the coefficients of x^3, x^2, x and the constant term gives $4 = b_2$, $b_1 - b_2 = 3$, $b_0 - b_1 = -6$ and $-b_0 = -1$, respectively, which are consistent if $b_1 = 7$. To find the two remaining roots we set $h(x) = 0$:

$$4x^2 + 7x + 1 = 0.$$

The roots of this quadratic equation are given by the standard formula as

$$\alpha_{1,2} = \frac{-7 \pm \sqrt{49 - 16}}{8}.$$

(c) When $k = 1$ (i.e. the original equation) the values of $g(x)$ at its turning points, $x = -1$ and $x = \frac{1}{2}$, are 4 and $-\frac{11}{4}$, respectively. Thus $g(x)$ can have up to 4 subtracted from it or up to $\frac{11}{4}$ added to it and still satisfy the condition for three (or, at the limit, two) distinct roots of $g(x) = 0$. It follows that for k outside the range $-5 \leq k \leq \frac{7}{4}$, $f(x) \; [= g(x) + 1 - k]$ has only one real root.

2.3 Investigate the properties of the polynomial equation

$$f(x) = x^7 + 5x^6 + x^4 - x^3 + x^2 - 2 = 0,$$

by proceeding as follows.

(a) By writing the fifth-degree polynomial appearing in the expression for $f'(x)$ in the form $7x^5 + 30x^4 + a(x - b)^2 + c$, show that there is in fact only one positive root of $f(x) = 0$.
(b) By evaluating $f(1)$, $f(0)$ and $f(-1)$, and by inspecting the form of $f(x)$ for negative values of x, determine what you can about the positions of the real roots of $f(x) = 0$.

(a) We start by finding the derivative of $f(x)$ and note that, because f contains no linear term, f' can be written as the product of x and a fifth-degree polynomial:

$$f(x) = x^7 + 5x^6 + x^4 - x^3 + x^2 - 2 = 0,$$
$$f'(x) = x(7x^5 + 30x^4 + 4x^2 - 3x + 2)$$
$$= x[7x^5 + 30x^4 + 4(x - \tfrac{3}{8})^2 - 4(\tfrac{3}{8})^2 + 2]$$
$$= x[7x^5 + 30x^4 + 4(x - \tfrac{3}{8})^2 + \tfrac{23}{16}].$$

Since, for positive x, every term in this last expression is necessarily positive, it follows that $f'(x)$ can have no zeros in the range $0 < x < \infty$. Consequently, $f(x)$ can have no turning points in that range and $f(x) = 0$ can have at most one root in the same range. However, $f(+\infty) = +\infty$ and $f(0) = -2 < 0$ and so $f(x) = 0$ has at least one root in $0 < x < \infty$. Consequently, it has exactly one root in the range.

(b) $f(1) = 5$, $f(0) = -2$ and $f(-1) = 5$, and so there is at least one root in each of the ranges $0 < x < 1$ and $-1 < x < 0$.

There is no simple systematic way to examine the form of a general polynomial function for the purpose of determining where its zeros lie, but it is sometimes helpful to group terms in the polynomial and determine how the sign of each group depends upon the range in which x lies. Here, grouping successive pairs of terms yields some information as follows:

$$x^7 + 5x^6 \quad \text{is positive for} \quad x > -5,$$
$$x^4 - x^3 \quad \text{is positive for} \quad x > 1 \quad \text{and} \quad x < 0,$$
$$x^2 - 2 \quad \text{is positive for} \quad x > \sqrt{2} \quad \text{and} \quad x < -\sqrt{2}.$$

Thus, all three terms are positive in the range(s) common to these, namely $-5 < x < -\sqrt{2}$ and $x > 1$. It follows that $f(x)$ is positive definite in these ranges and there can be no roots of $f(x) = 0$ within them. However, since $f(x)$ is negative for large negative x, there must be at least one root α with $\alpha < -5$.

2.5 Construct the quadratic equations that have the following pairs of roots:

(a) $-6, -3$; (b) $0, 4$; (c) $2, 2$; (d) $3 + 2i, 3 - 2i$, where $i^2 = -1$.

Starting in each case from the 'product of factors' form of the quadratic equation, $(x - \alpha_1)(x - \alpha_2) = 0$, we obtain:

(a) $\qquad (x + 6)(x + 3) = x^2 + 9x + 18 = 0;$

(b) $\qquad (x - 0)(x - 4) = x^2 - 4x = 0;$

(c) $\qquad (x - 2)(x - 2) = x^2 - 4x + 4 = 0;$

(d) $(x - 3 - 2i)(x - 3 + 2i) = x^2 + x(-3 - 2i - 3 + 2i) + (9 - 6i + 6i - 4i^2)$
$$= x^2 - 6x + 13 = 0.$$

2.7 Use the properties of the roots of a polynomial equation to prove that if the roots of $3x^3 - x^2 - 10x + 8 = 0$ are α_1, α_2 and α_3 then

(a) $\alpha_1^{-1} + \alpha_2^{-1} + \alpha_3^{-1} = 5/4$,
(b) $\alpha_1^2 + \alpha_2^2 + \alpha_3^2 = 61/9$,
(c) $\alpha_1^3 + \alpha_2^3 + \alpha_3^3 = -125/27$.
(d) Convince yourself that eliminating (say) α_2 and α_3 from (a), (b) and (c) does *not* give a simple explicit way of finding α_1.

In each case we must aim to write the given expression in terms of the combinations of the roots that can be directly expressed in terms of the coefficients of the original equation, e.g. $\sum_i \alpha_i$ or $\prod_i \alpha_i$.

(a) Write $\alpha_1^{-1} + \alpha_2^{-1} + \alpha_3^{-1}$ as

$$\frac{\alpha_2\alpha_3 + \alpha_1\alpha_3 + \alpha_2\alpha_1}{\alpha_1\alpha_2\alpha_3}.$$

Now, $\alpha_2\alpha_3 + \alpha_1\alpha_3 + \alpha_2\alpha_1 = (-10)/3$, whilst $\alpha_1\alpha_2\alpha_3 = -(8/3)$. And so $\alpha_1^{-1} + \alpha_2^{-1} + \alpha_3^{-1} = (-10/3)/(-8/3) = 5/4$.

(b) Write $\alpha_1^2 + \alpha_2^2 + \alpha_3^2$ as

$$(\alpha_1 + \alpha_2 + \alpha_3)^2 - 2(\alpha_1\alpha_2 + \alpha_2\alpha_3 + \alpha_3\alpha_1).$$

Now, $\alpha_1 + \alpha_2 + \alpha_3 = -(-1/3)$ and so $\alpha_1^2 + \alpha_2^2 + \alpha_3^2 = (1/3)^2 - 2(-10/3) = 61/9$.

(c) It is more difficult to see the required reformulation of this expression, but clearly taking the cube of $\sum_i \alpha_i$ will generate the required sum of cubes, as well as other terms (each containing at least two of the roots) that must be cancelled by appropriate products of symmetric combinations of the roots.

Proceeding in this way we find that $\alpha_1^3 + \alpha_2^3 + \alpha_3^3$ can be written as

$$(\alpha_1 + \alpha_2 + \alpha_3)^3 - 3(\alpha_1 + \alpha_2 + \alpha_3)(\alpha_1\alpha_2 + \alpha_2\alpha_3 + \alpha_3\alpha_1) + 3\alpha_1\alpha_2\alpha_3$$

$$= \left(\frac{1}{3}\right)^3 - 3\left(\frac{1}{3}\right)\left(\frac{-10}{3}\right) + 3\left(\frac{-8}{3}\right) = -\frac{125}{27}.$$

(d) No answer is available as it cannot be done. All manipulation is complicated and, at best, leads back to the original equation. Unfortunately, this is a 'proof by frustration', rather than one by contradiction.

2.9 The product of two numbers, α and β, is equal to λ times their sum, and their ratio is equal to μ times their sum. Find explicit expressions for α and β in terms of λ and μ.

We start from

$$\alpha\beta = \lambda(\alpha + \beta) \quad \text{and} \quad \frac{\alpha}{\beta} = \mu(\alpha + \beta).$$

Multiplying the two LHSs and the two RHSs together gives

$$\alpha^2 = \lambda\mu(\alpha + \beta)^2 \quad \Rightarrow \quad \beta^2 = \frac{\lambda^2(\alpha + \beta)^2}{\alpha^2} = \frac{\lambda}{\mu}, \quad \text{i.e. } \beta = \pm\sqrt{\frac{\lambda}{\mu}}.$$

Then, making α the subject of the first given equation, dividing both numerator and denominator by β, and finally substituting for β yields

$$\alpha = \frac{\lambda\beta}{\beta - \lambda} = \frac{\lambda}{1 \mp \sqrt{\lambda\mu}}.$$

Coordinate geometry

2.11 Determine the forms of the conic sections described by the following equations:

(a) $x^2 + y^2 + 6x + 8y = 0$;
(b) $9x^2 - 4y^2 - 54x - 16y + 29 = 0$;
(c) $2x^2 + 2y^2 + 5xy - 4x + y - 6 = 0$;
(d) $x^2 + y^2 + 2xy - 8x + 8y = 0$.

(a) $x^2 + y^2 + 6x + 8y = 0$. The coefficients of x^2 and y^2 are equal and there is no xy term; it follows that this must represent a circle. Rewriting the equation in standard circle form by 'completing the squares' in the terms that involve x and y, each variable treated separately, we obtain

$$(x + 3)^2 + (y + 4)^2 - (3^2 + 4^2) = 0.$$

The equation is therefore that of a circle of radius $\sqrt{3^2 + 4^2} = 5$ centred on $(-3, -4)$.

(b) $9x^2 - 4y^2 - 54x - 16y + 29 = 0$. This equation contains no xy term and so the centre of the curve will be at $(54/(2 \times 9), 16/[2 \times (-4)]) = (3, -2)$, and in standardised form the equation is

$$9(x - 3)^2 - 4(y + 2)^2 + 29 - 81 + 16 = 0,$$

or

$$\frac{(x - 3)^2}{4} - \frac{(y + 2)^2}{9} = 1.$$

The minus sign between the terms on the LHS implies that this conic section is a hyperbola with asymptotes (the form for large x and y and obtained by ignoring the constant on the RHS) given by $3(x - 3) = \pm 2(y + 2)$, i.e. lines of slope $\pm \frac{3}{2}$ passing through its 'centre' at $(3, -2)$.

(c) $2x^2 + 2y^2 + 5xy - 4x + y - 6 = 0$. As an xy term is present the equation cannot represent an ellipse or hyperbola in standard form. Whether it represents two straight lines can be most easily investigated by taking the lines in the form $a_i x + b_i y + 1 = 0$, $(i = 1, 2)$ and comparing the product $(a_1 x + b_1 y + 1)(a_2 x + b_2 y + 1)$ with $-\frac{1}{6}(2x^2 + 2y^2 + 5xy - 4x + y - 6)$. The comparison produces five equations which the four constants a_i, b_i, $(i = 1, 2)$ must satisfy:

$$a_1 a_2 = \frac{2}{-6}, \quad b_1 b_2 = \frac{2}{-6}, \quad a_1 + a_2 = \frac{-4}{-6}, \quad b_1 + b_2 = \frac{1}{-6}$$

and

$$a_1 b_2 + b_1 a_2 = \frac{5}{-6}.$$

Combining the first and third equations gives $3a_1^2 - 2a_1 - 1 = 0$, leading to a_1 and a_2 having the values 1 and $-\frac{1}{3}$, in either order. Similarly, combining the second and fourth equations gives $6b_1^2 + b_1 - 2 = 0$, leading to b_1 and b_2 having the values $\frac{1}{2}$ and $-\frac{2}{3}$, again in either order.

Either of the two combinations $(a_1 = -\frac{1}{3}, b_1 = -\frac{2}{3}, a_2 = 1, b_2 = \frac{1}{2})$ and $(a_1 = 1, b_1 = \frac{1}{2}, a_2 = -\frac{1}{3}, b_2 = -\frac{2}{3})$ also satisfies the fifth equation [note that the two alternative pairings do not do so]. That a consistent set can be found shows that the equation does indeed represent a pair of straight lines, $x + 2y - 3 = 0$ and $2x + y + 2 = 0$.

(d) $x^2 + y^2 + 2xy - 8x + 8y = 0$. We note that the first three terms can be written as a perfect square and so the equation can be rewritten as

$$(x + y)^2 = 8(x - y).$$

The two lines given by $x + y = 0$ and $x - y = 0$ are orthogonal and so the equation is of the form $u^2 = 4av$, which, for Cartesian coordinates u, v, represents a parabola passing through the origin, symmetric about the v-axis ($u = 0$) and defined for $v \geq 0$. Thus the original equation is that of a parabola, symmetric about the line $x + y = 0$, passing through the origin and defined in the region $x \geq y$.

2.13 A paraboloid of revolution whose focus is a distance a from its 'nose' rests symmetrically on the inside of a vertical cone $\rho = bz$, with their axes coincident. Find the distance between the nose of the paraboloid and the vertex of the cone.

Let the required distance be z_0. Since the distance of the focus of the paraboloid from its nose is a, the equation of the paraboloid is $x^2 + y^2 = \rho^2 = 4az$, referred to an origin coincident with the nose. However, as the nose is at $z = z_0$, referred to the given origin (at the vertex of the cone) the equation of the paraboloid is $\rho^2 = 4a(z - z_0)$.

On the ring of contact, the paraboloid and the cone must have a common value for ρ, and, since the cone is a tangent to the paraboloid on that ring, the relevant z-value must be a double root of the equation

$$b^2 z^2 = \rho^2 = 4a(z - z_0), \quad \text{i.e. of} \quad b^2 z^2 - 4az + 4az_0 = 0.$$

For this to be the case requires $(4a)^2 = 4(b^2)(4az_0)$, leading to $z_0 = a/b^2$.

2.15 The foci of the ellipse

$$\frac{x^2}{a^2} + \frac{y^2}{b^2} = 1$$

with eccentricity e are the two points $(-ae, 0)$ and $(ae, 0)$. Show that the sum of the distances from *any* point on the ellipse to the foci is $2a$. [The constancy of the sum of the distances from two fixed points can be used as an alternative defining property of an ellipse.]

Consider a general point (x, y) on the ellipse. Recalling that, when expressed in terms of the eccentricity of the ellipse, $b^2 = a^2(1 - e^2)$, the value of y^2 for any particular x is

$$y^2 = b^2 \left(1 - \frac{x^2}{a^2}\right) = (1 - e^2)(a^2 - x^2).$$

Now the distance from (x, y) to the focus at $(-ae, 0)$ is $[(x + ae)^2 + y^2]^{1/2}$, and that from the focus at $(ae, 0)$ is $[(x - ae)^2 + y^2]^{1/2}$. The sum of these distances is

$$\begin{aligned}
s &= [(x + ae)^2 + y^2]^{1/2} + [(x - ae)^2 + y^2]^{1/2} \\
&= [x^2 + 2aex + a^2 e^2 + (1 - e^2)(a^2 - x^2)]^{1/2} + \\
&\quad + [x^2 - 2aex + a^2 e^2 + (1 - e^2)(a^2 - x^2)]^{1/2} \\
&= [2aex + a^2 + e^2 x^2]^{1/2} + [-2aex + a^2 + e^2 x^2]^{1/2} \\
&= (a + ex) + (a - ex) = 2a,
\end{aligned}$$

irrespective of the value of x.

2.17 Describe and sketch the following parametrically defined curves.

(a) $x = t, \ y = t^{-1}$, (b) $x = \cos t, \ y = \sin t, \ z = t$,
(c) $x = t^3 - 3t, \ y = t^2 - 1$, (d) $x = a(t - \sin t), \ y = a(1 - \cos t)$,
(e) $x = a \cos^3 t, \ y = a \sin^3 t$, (f) $x = 4 \cos 3t, \ y = 3 \cos 2t$.

For the two-dimensional curves, see Figure 2.1.

(a) The x–y equation of the curve is simply $xy = 1$, which describes a rectangular hyperbola with the x- and y-axes as asymptotes.

(b) Since $x^2 + y^2 = 1$ for all t and z increases linearly with t, the curve is a spiral on a cylindrical surface of unit radius with its axis along the z-axis. The spiral has pitch 2π.

(c) The curve crosses the x-axis when $y = 0$, i.e. when $t = \pm 1$ at $x = \mp 2$. It crosses the y-axis at $t = 0$ and $t = \pm\sqrt{3}$, corresponding to $y = -1$ and $y = 2$ (twice). Asymptotically $y = x^{2/3}$.

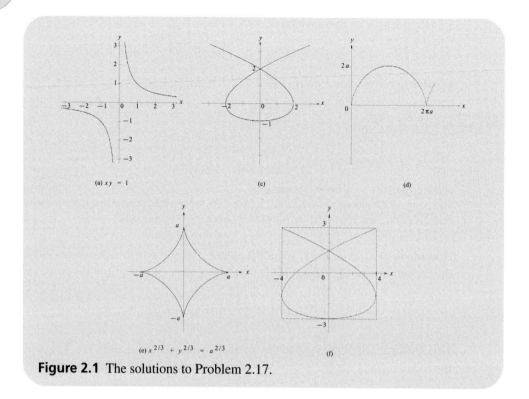

Figure 2.1 The solutions to Problem 2.17.

(d) This is the parametric form of a cycloid of 'amplitude' (maximum variation in y) equal to $2a$ and 'period' $2\pi a$. It cannot be written as a 'simple' equation $f(x, y) = 0$. At a cusp the tangent to the curve is vertical.

(e) In order to use the identity $\cos^2 t + \sin^2 t = 1$, we must rewrite the equation for x in the form $\cos^2 t = (x/a)^{2/3}$; similarly for the y equation. Then we have $(x/a)^{2/3} + (y/a)^{2/3} = 1$, giving the asteroid $x^{2/3} + y^{2/3} = a^{2/3}$.

(f) The curve is limited by the maximum and minimum values $x = \pm 4$, $y = \pm 3$. Because the two frequencies involved, 3 and 2, are integrally related, the curve will be closed. This particular curve reverses and then retraces its initial path after $t = \pi$.

2.19 Show that the locus of points in three-dimensional Cartesian space given by the parameterisation

$$x = au(3 - u^2), \qquad y = 3au^2, \qquad z = au(3 + u^2),$$

lies on the intersection of the surfaces $y^3 + 27axz - 81a^2y = 0$ and $y = \lambda(z - x)/(z + x)$, where λ is a constant you should determine.

Substituting the x-, y- and z-coordinates of the three-dimensional curve into the equation for the first surface produces

$$(3au^2)^3 + 27a[au(3 - u^2)][au(3 + u^2)] - 81a^2(3au^2) = 0,$$

$$27a^3u^6 + 27a^3u^2(9 - u^4) - 243a^3u^2 = 0,$$

which is identically satisfied. Therefore the curve lies in this first surface.

For it to lie on the second surface, $y = \lambda(z - x)/(z + x)$, as well, we require that

$$3au^2 = \lambda \frac{au(3 + u^2) - au(3 - u^2)}{au(3 + u^2) + au(3 - u^2)} = \lambda \frac{2au^3}{6au} = \lambda \frac{u^2}{3}.$$

This equation will be satisfied if $\lambda = 9a$. With this value for λ, the parameterised curve lies on the intersection of the two surfaces.

2.21 Identify the following curves, each given in plane polar coordinates.

(a) $\rho = 2a \sin \phi$,　　(b) $\rho = a + b\phi$,　　(c) $\rho \sin(\phi - \alpha) = p$,

where all symbols other than ρ and ϕ signify constants.

(a) Converting the equation from plane polar coordinates to Cartesian ones with $\rho^2 = (x^2 + y^2)^{1/2}$ and $\sin \phi = y/(x^2 + y^2)^{1/2}$, we obtain

$$(x^2 + y^2)^{1/2} = 2a \frac{y}{(x^2 + y^2)^{1/2}} \quad \Rightarrow \quad x^2 + y^2 = 2ay.$$

This can be rearranged as

$$(x - 0)^2 + (y - a)^2 = a^2,$$

showing that the curve is a circle of radius a centred on $(0, a)$.

The same conclusion can be reached by sketching a circle of diameter $2a$ that passes through the origin, and then constructing a right-angled triangle with the diameter through the origin (coincident with the y-axis) as its hypotenuse and the radius ρ as one of the other two sides.

(b) A direct interpretation using the definitions of ρ and ϕ shows that this curve is an equiangular spiral that starts at $(a, 0)$, and whose radius increases uniformly by $2\pi b$ for each turn of the spiral.

(c) In a rotated plane polar coordinate system (ρ, ϕ'), in which ϕ' is measured from the direction $\phi = \alpha$, the equation is $\rho \sin \phi' = p$. This is the equation of a line parallel to the direction $\phi' = 0$, and at a distance p from it. Therefore, in the original plane polar coordinates the curve is a straight line parallel to the line $\phi = \alpha$ and at a distance p from it.

2.23 Show that the equation of a standard ellipse with major axis $2a$ and eccentricity e can be expressed in the form

$$\rho = a \left(\frac{1 - e^2}{1 - e^2 \cos^2 \phi} \right)^{1/2},$$

using plane polar coordinates with their origin at the centre of the ellipse.

[*Note*: The usual plane polar description of an ellipse is $\rho = \ell(1 + e \cos \phi)^{-1}$, but this is referred to a coordinate system centred on a focus of the ellipse.]

In view of the form of the quoted expression, we aim to write everything in terms of ρ and $\cos\phi$. And so, setting $x = \rho\cos\phi$ and $y^2 = \rho^2(1 - \cos^2\phi)$, and recalling that $b^2 = a^2(1 - e^2)$, we have

$$\frac{\rho^2\cos^2\phi}{a^2} + \frac{\rho^2(1 - \cos^2\phi)}{a^2(1 - e^2)} = 1,$$

$$(1 - e^2)\rho^2\cos^2\phi + \rho^2(1 - \cos^2\phi) = a^2(1 - e^2),$$

$$\rho^2(1 - e^2\cos^2\phi) = a^2(1 - e^2),$$

$$\rho = a\left(\frac{1 - e^2}{1 - e^2\cos^2\phi}\right)^{1/2}.$$

Partial fractions

2.25 Resolve

(a) $\dfrac{2x + 1}{x^2 + 3x - 10}$,   (b) $\dfrac{4}{x^2 - 3x}$

into partial fractions using each of the following three methods:

(i) Expressing the supposed expansion in a form in which all terms have the same denominator and then equating coefficients of the various powers of x.

(ii) Substituting specific numerical values for x and solving the resulting simultaneous equations.

(iii) Evaluation of the fraction at each of the roots of its denominator, imagining a factored denominator with the factor corresponding to the root omitted – often known as the 'cover-up' method.

Verify that the decomposition obtained is independent of the method used.

(a) As the denominator factorises as $(x + 5)(x - 2)$, the partial fraction expansion must have the form

$$\frac{2x + 1}{x^2 + 3x - 10} = \frac{A}{x + 5} + \frac{B}{x - 2}.$$

(i)

$$\frac{A}{x + 5} + \frac{B}{x - 2} = \frac{x(A + B) + (5B - 2A)}{(x + 5)(x - 2)}.$$

Solving $A + B = 2$ and $-2A + 5B = 1$ gives $A = \frac{9}{7}$ and $B = \frac{5}{7}$.

(ii) Setting x equal to 0 and 1, say, gives the pair of equations

$$\frac{1}{-10} = \frac{A}{5} + \frac{B}{-2}; \quad \frac{3}{-6} = \frac{A}{6} + \frac{B}{-1},$$

$$-1 = 2A - 5B; \quad -3 = A - 6B,$$

with solution $A = \frac{9}{7}$ and $B = \frac{5}{7}$.

(iii)

$$A = \frac{2(-5) + 1}{-5 - 2} = \frac{9}{7}; \quad B = \frac{2(2) + 1}{2 + 5} = \frac{5}{7}.$$

All three methods give the same decomposition.

(b) Here the factorisation of the denominator is simply $x(x - 3)$ or, more formally, $(x - 0)(x - 3)$, and the expansion takes the form

$$\frac{4}{x^2 - 3x} = \frac{A}{x} + \frac{B}{x - 3}.$$

(i)

$$\frac{A}{x} + \frac{B}{x - 3} = \frac{x(A + B) - 3A}{(x - 0)(x - 3)}.$$

Solving $A + B = 0$ and $-3A = 4$ gives $A = -\frac{4}{3}$ and $B = \frac{4}{3}$.

(ii) Setting x equal to 1 and 2, say, gives the pair of equations

$$\frac{4}{-2} = \frac{A}{1} + \frac{B}{-2}; \quad \frac{4}{-2} = \frac{A}{2} + \frac{B}{-1},$$

$$-4 = 2A - B; \quad -4 = A - 2B,$$

with solution $A = -\frac{4}{3}$ and $B = \frac{4}{3}$.

(iii)

$$A = \frac{4}{0 - 3} = -\frac{4}{3}; \quad B = \frac{4}{3 - 0} = \frac{4}{3}.$$

Again, all three methods give the same decomposition.

2.27 Rearrange the following functions in partial fraction form:

(a) $\dfrac{x - 6}{x^3 - x^2 + 4x - 4}$, (b) $\dfrac{x^3 + 3x^2 + x + 19}{x^4 + 10x^2 + 9}$.

(a) For the function

$$f(x) = \frac{x - 6}{x^3 - x^2 + 4x - 4} = \frac{g(x)}{h(x)}$$

the first task is to factorise the denominator. By inspection, $h(1) = 0$ and so $x - 1$ is a factor of the denominator.

Write

$$x^3 - x^2 + 4x - 4 = (x - 1)(x^2 + b_1 x + b_0).$$

Equating coefficients: $-1 = b_1 - 1$, $4 = -b_1 + b_0$ and $-4 = -b_0$, giving $b_1 = 0$ and $b_0 = 4$. Thus,

$$f(x) = \frac{x - 6}{(x - 1)(x^2 + 4)}.$$

The factor $x^2 + 4$ cannot be factorised further without using complex numbers and so we include a term with this factor as the denominator, but 'at the price of' having a linear term, and not just a number, in the numerator.

$$f(x) = \frac{A}{x-1} + \frac{Bx + C}{x^2 + 4}$$

$$= \frac{Ax^2 + 4A + Bx^2 + Cx - Bx - C}{(x-1)(x^2+4)}.$$

Comparing the coefficients of the various powers of x in this numerator with those in the numerator of the original expression gives $A + B = 0$, $C - B = 1$ and $4A - C = -6$, which in turn yield $A = -1$, $B = 1$ and $C = 2$. Thus,

$$f(x) = -\frac{1}{x-1} + \frac{x+2}{x^2+4}.$$

(b) By inspection, the denominator of

$$\frac{x^3 + 3x^2 + x + 19}{x^4 + 10x^2 + 9}$$

factorises simply into $(x^2 + 9)(x^2 + 1)$, but neither factor can be broken down further. Thus, as in (a), we write

$$f(x) = \frac{Ax + B}{x^2 + 9} + \frac{Cx + D}{x^2 + 1}$$

$$= \frac{(A+C)x^3 + (B+D)x^2 + (A+9C)x + (B+9D)}{(x^2+9)(x^2+1)}.$$

Equating coefficients gives

$$A + C = 1,$$
$$B + D = 3,$$
$$A + 9C = 1,$$
$$B + 9D = 19.$$

From the first and third equations, $A = 1$ and $C = 0$. The second and fourth yield $B = 1$ and $D = 2$. Thus

$$f(x) = \frac{x+1}{x^2+9} + \frac{2}{x^2+1}.$$

Proof by induction and contradiction

2.29 Prove by induction that

$$\sum_{r=1}^{n} r = \tfrac{1}{2}n(n+1) \quad \text{and} \quad \sum_{r=1}^{n} r^3 = \tfrac{1}{4}n^2(n+1)^2.$$

To prove that

$$\sum_{r=1}^{n} r = \tfrac{1}{2}n(n+1),$$

assume that the result is valid for $n = N$ and consider

$$\sum_{r=1}^{N+1} r = \sum_{r=1}^{N} r + (N+1)$$

$$= \tfrac{1}{2}N(N+1) + (N+1), \quad \text{using the assumption,}$$

$$= (N+1)(\tfrac{1}{2}N+1)$$

$$= \tfrac{1}{2}(N+1)(N+2).$$

This is the same form as in the assumption except that N has been replaced by $N+1$; this shows that the result is valid for $n = N+1$ if it is valid for $n = N$. But the assumed result is trivially valid for $n = 1$ and is therefore valid for all n.

To prove that

$$\sum_{r=1}^{n} r^3 = \tfrac{1}{4}n^2(n+1)^2,$$

assume that the result is valid for $n = N$ and consider

$$\sum_{r=1}^{N+1} r^3 = \sum_{r=1}^{N} r^3 + (N+1)^3$$

$$= \tfrac{1}{4}N^2(N+1)^2 + (N+1)^3, \quad \text{using the assumption,}$$

$$= \tfrac{1}{4}(N+1)^2[N^2 + 4(N+1)]$$

$$= \tfrac{1}{4}(N+1)^2(N+2)^2.$$

This is the same form as in the assumption except that N has been replaced by $N+1$ and shows that the result is valid for $n = N+1$ if it is valid for $n = N$. But the assumed result is trivially valid for $n = 1$ and is therefore valid for all n.

2.31 Prove that $3^{2n} + 7$, where n is a non-negative integer, is divisible by 8.

As usual, we assume that the result is valid for $n = N$ and consider the expression with N replaced by $N + 1$:

$$3^{2(N+1)} + 7 = 3^{2N+2} + 7 + 3^{2N} - 3^{2N}$$

$$= (3^{2N} + 7) + 3^{2N}(9 - 1).$$

By the assumption, the first term on the RHS is divisible by 8; the second is clearly so. Thus $3^{2(N+1)} + 7$ is divisible by 8. This shows that the result is valid for $n = N + 1$ if it is valid for $n = N$. But the assumed result is trivially valid for $n = 0$ and is therefore valid for all n.

2.33 Establish the values of k for which the binomial coefficient ^pC_k is divisible by p when p is a prime number. Use your result and the method of induction to prove that $n^p - n$ is divisible by p for all integers n and all prime numbers p. Deduce that $n^5 - n$ is divisible by 30 for any integer n.

Since

$$^pC_k = \frac{p!}{k!(p-k)!},$$

its numerator will always contain a factor p. Therefore, the fraction will be divisible by p unless the denominator happens to contain a (cancelling) factor of p. Since p is prime, this latter factor cannot arise from the product of two or more terms in the denominator; nor can p have any factor that cancels with a term in the denominator. Thus, for cancellation to occur, either $k!$ or $(p-k)!$ must contain a term p; this can only happen for $k = p$ or $k = 0$; for all other values of k, ^pC_k will be divisible by p.

Assume that $n^p - n$ is divisible by prime number p for $n = N$. Clearly this is true for $N = 1$ and any p. Now, using the binomial expansion of $(N+1)^p$, consider

$$(N+1)^p - (N+1) = \sum_{k=0}^{p} {}^pC_k N^k - (N+1)$$

$$= 1 + \sum_{k=1}^{p-1} {}^pC_k N^k + N^p - N - 1.$$

But, as shown above, ^pC_k is divisible by p for all k in the range $1 \le k \le p-1$, and $N^p - N$ is divisible by p, by assumption. Thus $(N+1)^p - (N+1)$ is divisible by p if it is true that $N^p - N$ is divisible by p. Taking $N = 1$, for which, as noted above, the assumption is valid by inspection for any p, the result follows for all positive integers n and all primes p.

Now consider $f(n) = n^5 - n$. By the result just proved $f(n)$ is divisible by (prime number) 5. Further, $f(n) = n(n^4 - 1) = n(n^2 - 1)(n^2 + 1) = n(n-1)(n+1)(n^2 + 1)$. Thus the factorisation of $f(n)$ contains three consecutive integers; one of them must be divisible by 3 and at least one must be even and hence divisible by 2. Thus, $f(n)$ has the prime numbers 2, 3 and 5 as its divisors and must therefore be divisible by 30.

2.35 Prove, by the method of contradiction, that the equation

$$x^n + a_{n-1}x^{n-1} + \cdots + a_1 x + a_0 = 0,$$

in which all the coefficients a_i are integers, cannot have a rational root, unless that root is an integer. Deduce that any integral root must be a divisor of a_0 and hence find all rational roots of

(a) $x^4 + 6x^3 + 4x^2 + 5x + 4 = 0$,
(b) $x^4 + 5x^3 + 2x^2 - 10x + 6 = 0$.

Suppose that the equation has a rational root $x = p/q$, where integers p and q have no common factor and q is neither 0 nor 1. Then substituting the root and multiplying the resulting equation by q^{n-1} gives

$$\frac{p^n}{q} + a_{n-1}p^{n-1} + \cdots + a_1 pq^{n-2} + a_0 q^{n-1} = 0.$$

But the first term of this equation is not an integer (since p and q have no factor in common) whilst each of the remaining terms is a product of integers and is therefore an integer. Thus we have an integer equal to (minus) a non-integer. This is a contradiction and shows that it was wrong to suppose that the original equation has a rational non-integer root.

From the general properties of polynomial equations we have that the product of the roots of the equation $\sum_{i=0}^{n} b_i x^i = 0$ is $(-1)^n b_0/b_n$. For our original equation, $b_n = 1$ and $b_0 = a_0$. Consequently, the product of its roots is equal to the integral value $(-1)^n a_0$. Since there are no non-integral rational roots it follows that any integral root must be a divisor of a_0.

(a) $x^4 + 6x^3 + 4x^2 + 5x + 4 = 0$. This equation has integer coefficients and a leading coefficient equal to unity. We can thus apply the above result, which shows that its only possible rational roots are the six integers ± 1, ± 2 and ± 4. Of these, all positive values are impossible (since then every term would be positive) and trial and error will show that none of the negative values is a root either.

(b) $x^4 + 5x^3 + 2x^2 - 10x + 6 = 0$. In the same way as above, we deduce that for this equation the only possible rational roots are the eight values ± 1, ± 2, ± 3 and ± 6. Substituting each in turn shows that only $x = -3$ satisfies the equation.

Necessary and sufficient conditions

2.37 For the real variable x, show that a sufficient, but not necessary, condition for $f(x) = x(x+1) \times (2x+1)$ to be divisible by 6 is that x is an integer.

First suppose that x is an integer and consider $f(x)$ expressed as

$$f(x) = x(x+1)(2x+1) = x(x+1)(x+2) + x(x+1)(x-1).$$

Each term on the RHS consists of the product of three consecutive integers. In such a product one of the integers must divide by 3 and at least one of the other integers must be even. Thus each product separately divides by both 3 and 2, and hence by 6, and therefore so does their sum $f(x)$. Thus x being an integer *is a sufficient* condition for $f(x)$ to be divisible by 6.

That it is *not* a necessary condition can be shown by considering an equation of the form

$$f(x) = x(x+1)(2x+1) = 2x^3 + 3x^2 + x = 6m,$$

where m is an integer. As a specific counter-example consider the case $m = 4$. We note that $f(1) = 6$ whilst $f(2) = 30$. Thus there must be a root of the equation that lies strictly between the values 1 and 2, i.e a non-integer value of x that makes $f(x)$ equal to 24 and hence divisible by 6. This establishes the result that x being an integer is *not a necessary* condition for $f(x)$ to be divisible by 6.

2.39 The coefficients a_i in the polynomial $Q(x) = a_4 x^4 + a_3 x^3 + a_2 x^2 + a_1 x$ are all integers. Show that $Q(n)$ is divisible by 24 for all integers $n \geq 0$ if and only if all of the following conditions are satisfied:

(i) $2a_4 + a_3$ is divisible by 4;
(ii) $a_4 + a_2$ is divisible by 12;
(iii) $a_4 + a_3 + a_2 + a_1$ is divisible by 24.

This problem involves both proof by induction and proof of the 'if and only if' variety. Firstly, assume that the three conditions are satisfied:

$$2a_4 + a_3 = 4\alpha,$$

$$a_4 + a_2 = 12\beta,$$

$$a_4 + a_3 + a_2 + a_1 = 24\gamma,$$

where α, β and γ are integers. We now have to prove that $Q(n) = a_4 n^4 + a_3 n^3 + a_2 n^2 + a_1 n$ is divisible by 24 for all integers $n \geq 0$. It is clearly true for $n = 0$, and we assume that it is true for $n = N$ and that $Q(N) = 24m$ for some integer m. Now consider $Q(N + 1)$:

$$\begin{aligned}
Q(N + 1) &= a_4(N + 1)^4 + a_3(N + 1)^3 + a_2(N + 1)^2 + a_1(N + 1) \\
&= a_4 N^4 + a_3 N^3 + a_2 N^2 + a_1 N + 4a_4 N^3 + (6a_4 + 3a_3)N^2 \\
&\quad + (4a_4 + 3a_3 + 2a_2)N + (a_4 + a_3 + a_2 + a_1) \\
&= 24m + 4a_4 N^3 + 3(4\alpha)N^2 \\
&\quad + [4a_4 + (12\alpha - 6a_4) + (24\beta - 2a_4)]N + 24\gamma \\
&= 24(m + \gamma + \beta N) + 12\alpha N(N + 1) + 4a_4(N - 1)N(N + 1).
\end{aligned}$$

Now $N(N + 1)$ is the product of two consecutive integers and so one must be even and contain a factor of 2; likewise $(N - 1)N(N + 1)$, being the product of three consecutive integers, must contain both 2 and 3 as factors. Thus every term in the expression for $Q(N + 1)$ divides by 24 and so, therefore, does $Q(N + 1)$. Thus the proposal is true for $n = N + 1$ if it is true for $n = N$, and this, together with our observation for $n = 0$, completes the 'if' part of the proof.

Now suppose that $Q(n) = a_4 n^4 + a_3 n^3 + a_2 n^2 + a_1 n$ is divisible by 24 for all integers $n \geq 0$. Setting n equal to 1, 2 and 3 in turn, we have

$$a_4 + a_3 + a_2 + a_1 = 24p,$$

$$16a_4 + 8a_3 + 4a_2 + 2a_1 = 24q,$$

$$81a_4 + 27a_3 + 9a_2 + 3a_1 = 24r,$$

for some integers p, q and r. The first of these equations is condition (iii). The other conditions are established by combining the above equations as follows:

$$14a_4 + 6a_3 + 2a_2 = 24(q - 2p),$$
$$78a_4 + 24a_3 + 6a_2 = 24(r - 3p),$$
$$36a_4 + 6a_3 = 24(r - 3p - 3q + 6p),$$
$$22a_4 - 2a_2 = 24(r - 3p - 4q + 8p).$$

The two final equations show that $6a_4 + a_3$ is divisible by 4 and that $11a_4 - a_2$ is divisible by 12. But, if $6a_4 + a_3$ is divisible by 4 then so is $(6 - 4)a_4 + a_3$, i.e. $2a_4 + a_3$. Similarly, $11a_4 - a_2$ being divisible by 12 implies that $12a_4 - (11a_4 - a_2)$, i.e. $a_4 + a_2$, is also divisible by 12. Thus, conditions (i) and (ii) are established and the 'only if' part of the proof is complete.

3 Differential calculus

3.1 Obtain the following derivatives from first principles:

(a) the first derivative of $3x + 4$;
(b) the first, second and third derivatives of $x^2 + x$;
(c) the first derivative of $\sin 3x$.

(a) From the definition of the derivative as a limit, we have

$$f'(x) = \lim_{\Delta x \to 0} \frac{[3(x + \Delta x) + 4] - (3x + 4)}{\Delta x} = \lim_{\Delta x \to 0} \frac{3\Delta x}{\Delta x} = 3.$$

(b) These are calculated similarly, but using each calculated derivative as the input function for finding the next higher derivative.

$$f'(x) = \lim_{\Delta x \to 0} \frac{[(x + \Delta x)^2 + (x + \Delta x)] - (x^2 + x)}{\Delta x}$$

$$= \lim_{\Delta x \to 0} \frac{[(x^2 + 2x\,\Delta x + (\Delta x)^2) + (x + \Delta x)] - (x^2 + x)}{\Delta x}$$

$$= \lim_{\Delta x \to 0} \frac{[(2x\,\Delta x + (\Delta x)^2) + \Delta x]}{\Delta x}$$

$$= 2x + 1;$$

$$f''(x) = \lim_{\Delta x \to 0} \frac{[2(x + \Delta x) + 1] - (2x + 1)}{\Delta x} = \lim_{\Delta x \to 0} \frac{2\Delta x}{\Delta x} = 2;$$

$$f'''(x) = \lim_{\Delta x \to 0} \frac{2 - 2}{\Delta x} = 0.$$

(c) We use the expansion formula for $\sin(A + B)$ and then the series definitions of the sine and cosine functions to write $\cos \Delta x$ and $\sin \Delta x$ as series involving increasing powers of Δx.

$$f'(x) = \lim_{\Delta x \to 0} \frac{\sin 3(x + \Delta x) - \sin 3x}{\Delta x}$$

$$= \lim_{\Delta x \to 0} \frac{(\sin 3x \, \cos 3\Delta x + \cos 3x \, \sin 3\Delta x) - \sin 3x}{\Delta x}$$

$$= \lim_{\Delta x \to 0} \frac{\sin 3x \, (1 - \frac{(3\Delta x)^2}{2!} + \cdots) + \cos 3x \, (3\Delta x - \frac{(3\Delta x)^3}{3!} + \cdots) - \sin 3x}{\Delta x}$$

$$= \lim_{\Delta x \to 0} -\tfrac{1}{2}9\Delta x \sin 3x + 3\cos 3x - \tfrac{27}{6}(\Delta x)^2 \cos 3x + \cdots$$

$$= 3\cos 3x.$$

3.3 Find the first derivatives of

(a) $x^2 \exp x$,   (b) $2 \sin x \cos x$,   (c) $\sin 2x$,   (d) $x \sin ax$,

(e) $(e^{ax})(\sin ax) \tan^{-1} ax$,   (f) $\ln(x^a + x^{-a})$,   (g) $\ln(a^x + a^{-x})$,   (h) x^x.

(a) $x^2 \exp x$ is the product of two functions, both of which can be differentiated simply. We therefore apply the product rule and obtain

$$f'(x) = x^2 \frac{d(\exp x)}{dx} + \exp x \frac{d(x^2)}{dx} = x^2 \exp x + (2x) \exp x = (x^2 + 2x) \exp x.$$

(b) Again, the product rule is appropriate:

$$f'(x) = 2 \sin x \frac{d(\cos x)}{dx} + 2 \cos x \frac{d(\sin x)}{dx}$$

$$= 2 \sin x(- \sin x) + 2 \cos x(\cos x)$$

$$= 2(- \sin^2 x + \cos^2 x) = 2 \cos 2x.$$

(c) Rewriting the function as $f(x) = \sin u$, where $u(x) = 2x$, and using the chain rule:

$$f'(x) = \cos u \times \frac{du}{dx} = \cos u \times 2 = 2 \cos(2x).$$

We note that this is the same result as in part (b); this is not surprising, as the two functions to be differentiated are identical, i.e. $2 \sin x \cos x \equiv \sin 2x$.

(d) Once again, the product rule can be applied:

$$f'(x) = x \frac{d(\sin ax)}{dx} + \sin ax \frac{d(x)}{dx} = xa \cos ax + \sin ax \times 1 = \sin ax + ax \cos ax.$$

(e) This requires the product rule for three factors:

$$f'(x) = (e^{ax})(\sin ax) \frac{d(\tan^{-1} ax)}{dx} + (e^{ax})(\tan^{-1} ax) \frac{d(\sin ax)}{dx}$$

$$+ (\sin ax)(\tan^{-1} ax) \frac{d(e^{ax})}{dx}$$

$$= (e^{ax})(\sin ax) \left(\frac{a}{1 + a^2 x^2} \right) + (e^{ax})(\tan^{-1} ax)(a \cos ax)$$

$$+ (\sin ax)(\tan^{-1} ax)(ae^{ax})$$

$$= ae^{ax} \left[\frac{\sin ax}{1 + a^2 x^2} + (\tan^{-1} ax)(\cos ax + \sin ax) \right].$$

(f) Rewriting the function as $f(x) = \ln u$, where $u(x) = x^a + x^{-a}$, and using the chain rule:

$$f'(x) = \frac{1}{u} \times \frac{du}{dx} = \frac{1}{x^a + x^{-a}} \times (ax^{a-1} - ax^{-a-1}) = \frac{a(x^a - x^{-a})}{x(x^a + x^{-a})}.$$

(g) Using logarithmic differentiation and the chain rule as in (f):

$$f'(x) = \frac{1}{a^x + a^{-x}} \times (\ln a \, a^x - \ln a \, a^{-x}) = \frac{\ln a(a^x - a^{-x})}{a^x + a^{-x}}.$$

(h) In order to remove the independent variable x from the exponent in $y = x^x$, we first take logarithms and then differentiate implicitly:

$$y = x^x,$$

$$\ln y = x \ln x,$$

$$\frac{1}{y}\frac{dy}{dx} = \ln x + \frac{x}{x}, \quad \text{using the product rule,}$$

$$\frac{dy}{dx} = (1 + \ln x)x^x.$$

3.5 Use the result that $d[\,v(x)^{-1}\,]/dx = -v^{-2}dv/dx$ to find the first derivatives of (a) $(2x + 3)^{-3}$, (b) $\sec^2 x$, (c) $\operatorname{cosech}^3 3x$, (d) $1/\ln x$, (e) $1/[\sin^{-1}(x/a)]$.

(a) Writing $(2x + 3)^3$ as $v(x)$ and using the chain rule, we have

$$f'(x) = -\frac{1}{v^2}\frac{dv}{dx} = -\frac{1}{(2x + 3)^6}[\,3(2x + 3)^2\,(2)\,] = -\frac{6}{(2x + 3)^4}.$$

(b) Writing $\cos^2 x$ as $v(x)$, we have

$$f'(x) = -\frac{1}{v^2}\frac{dv}{dx} = -\frac{1}{\cos^4 x}[\,2\cos x(-\sin x)\,] = 2\sec^2 x \tan x.$$

(c) Writing $\sinh^3 3x$ as $v(x)$, we have

$$f'(x) = -\frac{1}{v^2}\frac{dv}{dx} = -\frac{1}{\sinh^6 3x}[\,3\sinh^2 3x(\cosh 3x)(3)\,]$$

$$= -9\operatorname{cosech}^3 3x \coth 3x.$$

(d) Writing $\ln x$ as $v(x)$, we have

$$f'(x) = -\frac{1}{v^2}\frac{dv}{dx} = -\frac{1}{(\ln x)^2}\frac{1}{x} = -\frac{1}{x \ln^2 x}.$$

(e) Writing $\sin^{-1}(x/a)$ as $v(x)$, we have

$$f'(x) = -\frac{1}{v^2}\frac{dv}{dx} = -\frac{1}{[\,\sin^{-1}(x/a)\,]^2}\frac{1}{\sqrt{a^2 - x^2}}.$$

3.7 Find the first derivative of

$$f(x) = \frac{x}{(x^2 + a^2)^{1/2}}$$

by making the substitution $x = a \tan \theta$. Show that $f(x) = g(\theta) = \sin \theta$ and then use the chain rule to obtain the derivative.

Making the substitution $x = a \tan \theta$ with $dx/d\theta = a \sec^2 \theta$, we obtain

$$f(x) = \frac{x}{(x^2 + a^2)^{1/2}} = \frac{a \tan \theta}{(a^2 \tan^2 \theta + a^2)^{1/2}} = \frac{\tan \theta}{\sec \theta} = \sin \theta = g(\theta).$$

Now, using the chain rule,

$$\frac{df}{dx} = \frac{df}{d\theta} \frac{d\theta}{dx} = \cos \theta \, \frac{1}{a \sec^2 \theta} = \frac{1}{a \sec^3 \theta} = \frac{1}{a(1 + \tan^2 \theta)^{3/2}}$$

$$= \frac{1}{a(1 + (x/a)^2)^{3/2}} = \frac{a^2}{(a^2 + x^2)^{3/2}}.$$

3.9 Find dy/dx if $x = (t - 2)/(t + 2)$ and $y = 2t/(t + 1)$ for $-\infty < t < \infty$. Show that it is always non-negative and make use of this result in sketching the curve of y as a function of x.

We calculate dy/dx as $dy/dt \div dx/dt$:

$$\frac{dy}{dt} = \frac{(t + 1)2 - 2t(1)}{(t + 1)^2} = \frac{2}{(t + 1)^2},$$

$$\frac{dx}{dt} = \frac{(t + 2)(1) - (t - 2)(1)}{(t + 2)^2} = \frac{4}{(t + 2)^2},$$

$$\Rightarrow \quad \frac{dy}{dx} = \frac{2}{(t + 1)^2} \div \frac{4}{(t + 2)^2} = \frac{(t + 2)^2}{2(t + 1)^2},$$

which is clearly positive for all t.

By evaluating x and y for a range of values of t and recalling that its slope is always positive, the curve can be plotted as in Figure 3.1. Alternatively, we may eliminate t using

$$t = \frac{2x + 2}{1 - x} \quad \text{and} \quad t = \frac{y}{2 - y},$$

to obtain the equation of the curve in x–y coordinates as

$$2(x + 1)(2 - y) = y(1 - x),$$

$$xy - 4x + 3y - 4 = 0,$$

$$(x + 3)(y - 4) = 4 - 12 = -8.$$

This shows that the curve is a rectangular hyperbola in the second and fourth quadrants with asymptotes, parallel to the x- and y-axes, passing through $(-3, 4)$.

3.11 Find the second derivative of $y(x) = \cos[(\pi/2) - ax]$. Now set $a = 1$ and verify that the result is the same as that obtained by first setting $a = 1$ and simplifying $y(x)$ before differentiating.

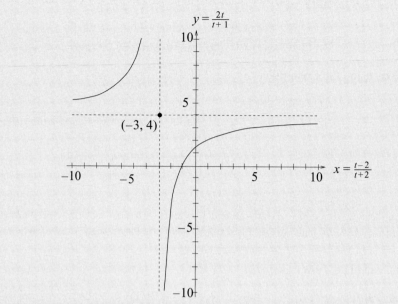

Figure 3.1 The solution to Problem 3.9.

We use the chain rule at each stage and, either finally or initially, the equality of $\cos(\frac{1}{2}\pi - \theta)$ and $\sin\theta$:

$$y(x) = \cos\left(\frac{\pi}{2} - ax\right),$$

$$y'(x) = a\sin\left(\frac{\pi}{2} - ax\right),$$

$$y''(x) = -a^2\cos\left(\frac{\pi}{2} - ax\right).$$

For $a = 1$, $\quad y''(x) = -\cos\left(\frac{\pi}{2} - x\right) = -\sin x.$

Setting $a = 1$ initially, gives $y = \cos(\frac{1}{2}\pi - x) = \sin x$. Hence $y' = \cos x$ and $y'' = -\sin x$, yielding the same result as before.

3.13 Show by differentiation and substitution that the differential equation

$$4x^2\frac{d^2y}{dx^2} - 4x\frac{dy}{dx} + (4x^2 + 3)y = 0$$

has a solution of the form $y(x) = x^n \sin x$, and find the value of n.

The solution plan is to calculate the derivatives as functions of n and x and then, after substitution, require that the equation is identically satisfied for all x. This will impose conditions on n.

We have, by successive differentiation or by the use of Leibnitz's theorem, that

$$y(x) = x^n \sin x,$$

$$y'(x) = nx^{n-1} \sin x + x^n \cos x,$$

$$y''(x) = n(n-1)x^{n-2} \sin x + 2nx^{n-1} \cos x - x^n \sin x.$$

Substituting these into

$$4x^2 \frac{d^2 y}{dx^2} - 4x \frac{dy}{dx} + (4x^2 + 3)y = 0$$

gives

$$(4n^2 - 4n - 4n + 3)x^n \sin x + (-4 + 4)x^{n+2} \sin x + (8n - 4)x^{n+1} \cos x = 0.$$

For this to be true for all x, both $4n^2 - 8n + 3 = (2n - 3)(2n - 1) = 0$ and $8n - 4 = 0$ have to be satisfied. If $n = \frac{1}{2}$, they *are* both satisfied, thus establishing $y(x) = x^{1/2} \sin x$ as a solution of the given equation.

3.15 Show that the lowest value taken by the function $3x^4 + 4x^3 - 12x^2 + 6$ is -26.

We need to calculate the first and second derivatives of the function in order to establish the positions and natures of its turning points:

$$y(x) = 3x^4 + 4x^3 - 12x^2 + 6,$$

$$y'(x) = 12x^3 + 12x^2 - 24x,$$

$$y''(x) = 36x^2 + 24x - 24.$$

Setting $y'(x) = 0$ gives $x(x + 2)(x - 1) = 0$ with roots 0, 1 and -2. The corresponding values of $y''(x)$ are -24, 36 and 72.

Since $y(\pm\infty) = \infty$, the lowest value of y is that corresponding to the lowest minimum, which can only be at $x = 1$ or $x = -2$, as y'' must be positive at a minimum. The values of $y(x)$ at these two points are $y(1) = 1$ and $y(-2) = -26$, and so the lowest value taken is -26.

3.17 Show that $y(x) = xa^{2x} \exp x^2$ has no stationary points other than $x = 0$, if

$$\exp(-\sqrt{2}) < a < \exp(\sqrt{2}).$$

Since the logarithm of a variable varies monotonically with the variable, the stationary points of the logarithm of a function of x occur at the same values of x as the stationary points of the function. As x appears as an exponent in the given function, we take logarithms before differentiating and obtain

$$\ln y = \ln x + 2x \ln a + x^2,$$

$$\frac{1}{y} \frac{dy}{dx} = \frac{1}{x} + 2 \ln a + 2x.$$

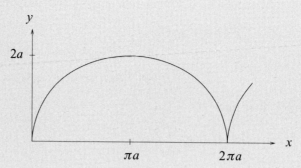

Figure 3.2 The solution to Problem 3.19.

For a stationary point $dy/dx = 0$. Except at $x = 0$ (where y is also 0), this equation reduces to

$$2x^2 + 2x \ln a + 1 = 0.$$

This quadratic equation has no real roots for x if $4(\ln a)^2 < 4 \times 2 \times 1$, i.e. $|\ln a| < \sqrt{2}$; a result that can also be written as $\exp(-\sqrt{2}) < a < \exp(\sqrt{2})$.

3.19 The parametric equations for the motion of a charged particle released from rest in electric and magnetic fields at right angles to each other take the forms

$$x = a(\theta - \sin\theta), \qquad y = a(1 - \cos\theta).$$

Show that the tangent to the curve has slope $\cot(\theta/2)$. Use this result at a few calculated values of x and y to sketch the form of the particle's trajectory.

With the given parameterisation,

$$\frac{dx}{d\theta} = a - a\cos\theta,$$

$$\frac{dy}{d\theta} = a\sin\theta,$$

$$\Rightarrow \quad \frac{dy}{dx} = \frac{dy}{d\theta}\frac{d\theta}{dx} = \frac{\sin\theta}{1 - \cos\theta} = \frac{2\sin\frac{1}{2}\theta\cos\frac{1}{2}\theta}{2\sin^2\frac{1}{2}\theta} = \cot\frac{1}{2}\theta.$$

Clearly, $y = 0$ whenever $\theta = 2n\pi$ with n an integer; dy/dx becomes infinite at the same points. The slope is zero whenever $\theta = (2n + 1)\pi$ and the value of y is then $2a$. These results are plotted in Figure 3.2.

3.21 The curve whose equation is $x^{2/3} + y^{2/3} = a^{2/3}$ for positive x and y and which is completed by its symmetric reflections in both axes is known as an astroid. Sketch it and show that its radius of curvature in the first quadrant is $3(axy)^{1/3}$.

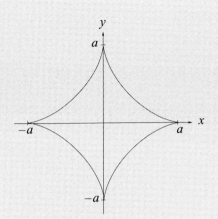

Figure 3.3 The astroid discussed in Problem 3.21.

For the astroid curve (see Figure 3.3) and its first derivative in the first quadrant, where all fractional roots are positive, we have

$$x^{2/3} + y^{2/3} = a^{2/3},$$

$$\frac{2}{3x^{1/3}} + \frac{2}{3y^{1/3}}\frac{dy}{dx} = 0,$$

$$\Rightarrow \quad \frac{dy}{dx} = -\left(\frac{y}{x}\right)^{1/3}.$$

Differentiating again:

$$\frac{d^2y}{dx^2} = -\frac{1}{3}\left(\frac{y}{x}\right)^{-2/3}\left[\frac{-x(\frac{y}{x})^{1/3} - y}{x^2}\right]$$

$$= \frac{1}{3}(x^{-2/3}y^{-1/3} + x^{-4/3}y^{1/3})$$

$$= \frac{1}{3}y^{-1/3}x^{-4/3}(x^{2/3} + y^{2/3})$$

$$= \frac{1}{3}y^{-1/3}x^{-4/3}a^{2/3}.$$

Hence, the radius of curvature is

$$\rho = \frac{\left[1 + \left(\frac{dy}{dx}\right)^2\right]^{3/2}}{\frac{d^2y}{dx^2}} = \frac{\left[1 + \left(\frac{y}{x}\right)^{2/3}\right]^{3/2}}{\frac{1}{3}y^{-1/3}x^{-4/3}a^{2/3}}$$

$$= 3(x^{2/3} + y^{2/3})^{3/2}x^{1/3}y^{1/3}a^{-2/3} = 3a^{1/3}x^{1/3}y^{1/3},$$

as stated in the question.

3.23 Use Leibnitz's theorem to find

(a) the second derivative of $\cos x \sin 2x$,
(b) the third derivative of $\sin x \ln x$,
(c) the fourth derivative of $(2x^3 + 3x^2 + x + 2)e^{2x}$.

Leibnitz's theorem states that if $y(x) = u(x)v(x)$ and the rth derivative of a function $f(x)$ is denoted by $f^{(r)}$, then

$$y^{(n)} = \sum_{k=0}^{n} {}^nC_k\, u^{(k)}\, v^{(n-k)}.$$

So:

(a) $\dfrac{d^2(\cos x \sin 2x)}{dx^2} = (-\cos x)(\sin 2x) + 2(-\sin x)(2\cos 2x) + (\cos x)(-4\sin 2x)$

$$= -5\cos x \sin 2x - 4\sin x \cos 2x$$

$$= 2\sin x[-5\cos^2 x - 2(2\cos^2 x - 1)]$$

$$= 2\sin x(2 - 9\cos^2 x).$$

(b) $\dfrac{d^3(\sin x \ln x)}{dx^3} = (-\cos x)(\ln x) + 3(-\sin x)(x^{-1}) + 3(\cos x)(-x^{-2}) + (\sin x)(2x^{-3})$

$$= (2x^{-3} - 3x^{-1})\sin x - (3x^{-2} + \ln x)\cos x.$$

(c) We note that the nth derivative of e^{2x} is $2^n e^{2x}$ and that the fourth derivative of a cubic polynomial is zero. And so:

$\dfrac{d^4[(2x^3 + 3x^2 + x + 2)e^{2x}]}{dx^4} = (0)(e^{2x}) + 4(12)(2e^{2x}) + 6(12x + 6)(4e^{2x})$

$$+ 4(6x^2 + 6x + 1)(8e^{2x}) + (2x^3 + 3x^2 + x + 2)(16e^{2x})$$

$$= 16(2x^3 + 15x^2 + 31x + 19)e^{2x}.$$

3.25 Use the properties of functions at their turning points to do the following.

(a) By considering its properties near $x = 1$, show that $f(x) = 5x^4 - 11x^3 + 26x^2 - 44x + 24$ takes negative values for some range of x.
(b) Show that $f(x) = \tan x - x$ cannot be negative for $0 \le x < \pi/2$, and deduce that $g(x) = x^{-1}\sin x$ decreases monotonically in the same range.

(a) We begin by evaluating $f(1)$ and find that $f(1) = 5 - 11 + 26 - 44 + 24 = 0$. This suggests that $f(x)$ will be positive on one side of $x = 1$ and negative on the other. However, to be sure of this we need to establish that $x = 1$ is *not* a turning point of $f(x)$. To do this we calculate its derivative there:

$$f(x) = 5x^4 - 11x^3 + 26x^2 - 44x + 24,$$

$$f'(x) = 20x^3 - 33x^2 + 52x - 44,$$

$$f'(1) = 20 - 33 + 52 - 44 = -5 \neq 0.$$

So $f'(1)$ is negative and f is decreasing at this point, where its value is 0. Therefore $f(x)$ must be negative in the range $1 < x < \alpha$ for some $\alpha > 1$.

(b) The function $f(x) = \tan x - x$ is differentiable in the range $0 \leq x < \pi/2$, and $f'(x) = \sec^2 x - 1 = \tan^2 x$, which is > 0 for all x in the range; taken together with $f(0) = 0$, this establishes the result.

For $g(x) = (\sin x)/x$, the rule for differentiating quotients gives

$$g'(x) = \frac{x \cos x - \sin x}{x^2} = -\frac{\cos x(\tan x - x)}{x^2}.$$

The term in parentheses cannot be negative in the range $0 \leq x < \pi/2$, and in the same range $\cos x > 0$. Thus $g'(x)$ is never positive in the range and $g(x)$ decreases monotonically [from its value of $g(0) = 1$].

3.27 By applying Rolle's theorem to $x^n \sin nx$, where n is an arbitrary positive integer, show that $\tan nx + x = 0$ has a solution α_1 with $0 < \alpha_1 < \pi/n$. Apply the theorem a second time to obtain the nonsensical result that there is a real α_2 in $0 < \alpha_2 < \pi/n$ such that $\cos^2(n\alpha_2) = -n$. Explain why this incorrect result arises.

Clearly, the function $f(x) = x^n \sin nx$ has zeros at $x = 0$ and $x = \pi/n$. Therefore, by Rolle's theorem, its derivative,

$$f'(x) = nx^{n-1} \sin nx + nx^n \cos nx,$$

must have a zero in the range $0 < x < \pi/n$. But, since $x \neq 0$ and $n \neq 0$, this is equivalent to a root α_1 of $\tan nx + x = 0$ in the same range. To obtain this result we have divided $f'(x) = 0$ through by $\cos nx$; this is allowed, since $x = \pi/(2n)$, the value that makes $\cos nx = 0$, is not a solution of $f'(x) = 0$.

We now note that $g(x) = \tan nx + x$ has zeros at $x = 0$ and $x = \alpha_1$. Applying Rolle's theorem again (blindly) then shows that $g'(x) = n \sec^2 nx + 1$ has a zero α_2 in the range $0 < \alpha_2 < \alpha_1 < \pi/n$, with $\cos^2(n\alpha_2) = -n$.

The false result arises because $\tan nx$ is not differentiable at $x = \pi/(2n)$, which lies in the range $0 < x < \pi/n$, and so the conditions for applying Rolle's theorem are not satisfied.

3.29 For the function $y(x) = x^2 \exp(-x)$ obtain a simple relationship between y and dy/dx and then, by applying Leibnitz's theorem, prove that

$$xy^{(n+1)} + (n + x - 2)y^{(n)} + ny^{(n-1)} = 0.$$

The required function and its first derivative are

$$y(x) = x^2 e^{-x},$$

$$y'(x) = 2xe^{-x} - x^2 e^{-x}$$

$$= 2xe^{-x} - y.$$

Multiplying through by a factor x will enable us to express the first term on the RHS in terms of y and obtain

$$xy' = 2y - xy.$$

Now we apply Leibnitz's theorem to obtain the nth derivatives of both sides of this last equation, noting that the only non-zero derivative of x is the first derivative. We obtain

$$xy^{(n+1)} + n(1)y^{(n)} = 2y^{(n)} - [xy^{(n)} + n(1)y^{(n-1)}],$$

which can be rearranged as

$$xy^{(n+1)} + (n + x - 2)y^{(n)} + ny^{(n-1)} = 0,$$

thus completing the proof.

3.31 Show that the curve $x^3 + y^3 - 12x - 8y - 16 = 0$ touches the x-axis.

We first find an expression for the slope of the curve as a function of x and y. From

$$x^3 + y^3 - 12x - 8y - 16 = 0$$

we obtain, by implicit differentiation, that

$$3x^2 + 3y^2 y' - 12 - 8y' = 0 \quad \Rightarrow \quad y' = \frac{3x^2 - 12}{8 - 3y^2}.$$

Clearly $y' = 0$ at $x = \pm 2$. At $x = 2$,

$$8 + y^3 - 24 - 8y - 16 = 0 \quad \Rightarrow \quad y \neq 0.$$

However, at $x = -2$,

$$-8 + y^3 + 24 - 8y - 16 = 0, \quad \text{with one solution } y = 0.$$

Thus the point $(-2, 0)$ lies on the curve and $y' = 0$ there. It follows that the curve touches the x-axis at that point.

3.33 Investigate the properties of the following functions and in each case make a sketchgraph incorporating the features you have identified.

(a) $f(x) = (x^2 + 4x + 2)/[x(x + 2)]$.
(b) $f(x) = [x(x^2 + 2x + 2)]/(x + 2)$.
(c) $f(x) = 1 - e^{-x/3}(\frac{1}{6} \sin 2x + \cos 2x)$.

(a) The function

$$f(x) = \frac{x^2 + 4x + 2}{x(x + 2)}$$

has zeros where

$$x^2 + 4x + 2 = 0, \quad \text{i.e. at } x = -2 \pm \sqrt{2^2 - (1)(2)} = -2 \pm \sqrt{2}.$$

Any turning points will be given (using the derivative of a quotient) by

$$x(x + 2)(2x + 4) - (x^2 + 4x + 2)(2x + 2) = 0 \quad \Rightarrow \quad 2x^2 + 4x + 4 = 0.$$

Since $(4)^2 < 4(2)(4)$ this quadratic equation has no real roots; consequently, $f(x)$ has no turning points.

The function will have vertical asymptotes at $x = 0$ and $x = -2$, and have the same sign as x near $x = 0$. Near $x = -2$, $f(x) \approx -2/(-2)(x + 2)$ and so is negative to the left of it.

Clearly, $f(x) \to 1$ as $x \to \pm\infty$. To determine in which way, we write $f(x)$ in the form

$$1 + \frac{x^2 + 4x + 2 - x(x + 2)}{x(x + 2)} = 1 + \frac{2x + 2}{x(x + 2)}.$$

From this form we see that $f(x) \to 1$ from above as $x \to +\infty$ and from below as $x \to -\infty$.

A sketchgraph is shown in Figure 3.4(a).

(b) Proceeding in a similar manner to that used in part (a), $f(x)$ has a zero only at $x = 0$ (since $x^2 + 2x + 2 = 0$ has no real roots). Any turning points will be given by

$$(x + 2)(3x^2 + 4x + 2) - (x^3 + 2x^2 + 2x)(1) = 0$$

$$\Rightarrow \quad g(x) = 2x^3 + 8x^2 + 8x + 4 = 0.$$

This cubic equation will have either one or three roots. To help decide which, we calculate $g'(x) = 6x^2 + 16x + 8 = (3x + 2)(2x + 4)$. Thus $g'(x) = 0$ has roots $x = -2$ and $x = -2/3$; it follows that $g(x)$ has turning points at the same two values of x. Now $g(-2) = 4$ and $g(-2/3) \approx 13.5$; as they have the same sign, $g(x)$ has no zeros between the two x-values. We conclude that $f(x)$ has only one turning point. Noting that $g(-3)$ is negative, whereas $g(-2)$ is positive, locates the turning point of f in this range [actually, $x \approx -2.85$].

The function will have vertical asymptote at $x = -2$. Near $x = -2$, $f(x) \approx -2(2)/(x + 2)$ and so is large and positive to the left of it.

We investigate the behaviour at $\pm\infty$ by writing $f(x)$ as

$$f(x) = \frac{x^3 + 2x^2 + 2x}{x + 2} = x^2 + \frac{2x}{x + 2} = x^2 + 2 - \frac{4}{x + 2}.$$

This shows that $f(x)$ is asymptotic to the curve $y = x^2 + 2$ approaching it from above as $x \to -\infty$ and from below as $x \to +\infty$.

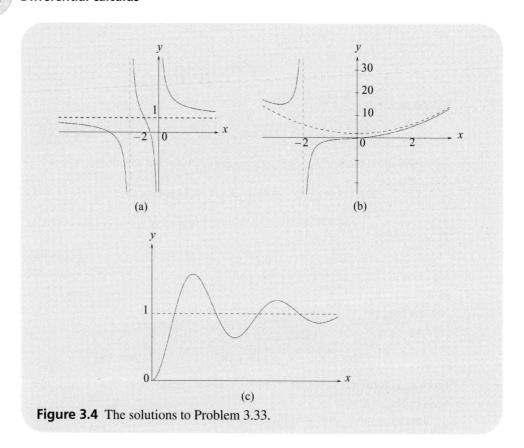

Figure 3.4 The solutions to Problem 3.33.

A sketchgraph is shown in Figure 3.4(b).

(c) This is clearly a damped oscillation of frequency 2 and an amplitude reduction in the oscillation of e^{-1} for every three units of increase in x.

As $x \to \infty$, $f(x) \to 1$, with the damped oscillations centred on that value. Further, $f(0) = 1 - e^{0}[0 + 1] = 0$, and so the curve includes an initial increase from zero to about unity; $f(x)$ first crosses $f = 1$ when $\tan 2x = -6$, i.e. $x \approx 0.87$.

A sketchgraph is shown in Figure 3.4(c).

4 Integral calculus

4.1 Find, by inspection, the indefinite integrals of

(a) $7x^6$; (b) $e^{3x} + e^{-3x}$; (c) $\cot 3x$; (d) $\sin x \sin 2x$; (e) $\cos x \sin 2x$;
(f) $(a - 2x)^{-1}$; (g) $(4 + x^2)^{-1}$; (h) $(4 - x^2)^{-1/2}$; (i) $x(4 + x^2)^{-1}$.

Since the solutions are 'by inspection', most of them consist of the answer only; in each case c is the constant of integration.

(a) $x^7 + c$.

(b) $\frac{1}{3}(e^{3x} - e^{-3x}) + c$.

(c) Since $\cot 3x = \cos 3x / \sin 3x$, the integral is $\frac{1}{3}\ln(\sin 3x) + c$.

(d) Since $\sin x \sin 2x = 2\sin^2 x \cos x$, the integral is $\frac{2}{3}\sin^3 x + c$.

(e) Since $\cos x \sin 2x = 2\sin x \cos^2 x$, the integral is $-\frac{2}{3}\cos^3 x + c$.

(f) $-\frac{1}{2}\ln(a - 2x) + c = \ln[(a - 2x)^{-1/2}] + c$.

(g) $\frac{1}{2}\tan^{-1}(x/2) + c$.

(h) $\sin^{-1}(x/2) + c$.

(i) $\frac{1}{2}\ln(4 + x^2) + c = \ln(4 + x^2)^{1/2} + c$.

4.3 Find the indefinite integrals J of the following ratios of polynomials:

(a) $(x + 3)/(x^2 + x - 2)$;
(b) $(x^3 + 5x^2 + 8x + 12)/(2x^2 + 10x + 12)$;
(c) $(3x^2 + 20x + 28)/(x^2 + 6x + 9)$;
(d) $x^3/(a^8 + x^8)$.

(a) We first need to express the ratio in partial fractions:

$$\frac{x + 3}{x^2 + x - 2} = \frac{x + 3}{(x + 2)(x - 1)} = \frac{A}{x + 2} + \frac{B}{x - 1}.$$

Using any of the methods employed in Problem 2.25, we obtain the unknown coefficients as $A = -\frac{1}{3}$ and $B = \frac{4}{3}$. Thus,

$$\int \frac{x + 3}{x^2 + x - 2}\, dx = \int \frac{-1}{3(x + 2)}\, dx + \int \frac{4}{3(x - 1)}\, dx$$

$$= -\frac{1}{3}\ln(x + 2) + \frac{4}{3}\ln(x - 1) + c$$

$$= \frac{1}{3}\ln\frac{(x - 1)^4}{x + 2} + c.$$

(b) As the numerator is of higher degree than the denominator, we need to divide the numerator by the denominator and express the remainder in partial fractions before starting any integration:

$$x^3 + 5x^2 + 8x + 12 = (\tfrac{1}{2}x + a_0)(2x^2 + 10x + 12) + (b_1 x + b_0)$$
$$= x^3 + (2a_0 + 5)x^2 + (10a_0 + 6 + b_1)x + (12a_0 + b_0),$$

yielding $a_0 = 0$, $b_1 = 2$ and $b_0 = 12$. Now, expressed as partial fractions,

$$\frac{2x + 12}{2x^2 + 10x + 12} = \frac{x + 6}{(x + 2)(x + 3)} = \frac{4}{x + 2} + \frac{-3}{x + 3},$$

where, again, we have used one of the three methods available for determining coefficients in partial fraction expansions. Thus,

$$\int \frac{x^3 + 5x^2 + 8x + 12}{2x^2 + 10x + 12}\, dx = \int \left(\frac{1}{2}x + \frac{4}{x + 2} - \frac{3}{x + 3} \right) dx$$
$$= \tfrac{1}{4}x^2 + 4 \ln(x + 2) - 3 \ln(x + 3) + c.$$

(c) By inspection,

$$3x^2 + 20x + 28 = 3(x^2 + 6x + 9) + 2x + 1.$$

Expressing the remainder after dividing through by $x^2 + 6x + 9$ in partial fractions, and noting that the denominator has a double factor, we obtain

$$\frac{2x + 1}{x^2 + 6x + 9} = \frac{A}{(x + 3)^2} + \frac{B}{x + 3},$$

where $B(x + 3) + A = 2x + 1$. This requires that $B = 2$ and $A = -5$. Thus,

$$\int \frac{3x^2 + 20x + 28}{x^2 + 6x + 9}\, dx = \int \left[3 + \frac{2}{x + 3} - \frac{5}{(x + 3)^2} \right] dx$$
$$= 3x + 2 \ln(x + 3) + \frac{5}{x + 3} + c.$$

(d) Noting the form of the numerator, we set $x^4 = u$ with $4x^3\, dx = du$. Then,

$$\int \frac{x^3}{a^8 + x^8}\, dx = \int \frac{1}{4(a^8 + u^2)}\, du$$
$$= \frac{1}{4a^4} \tan^{-1} \frac{u}{a^4} + c = \frac{1}{4a^4} \tan^{-1} \left(\frac{x^4}{a^4} \right) + c.$$

4.5 Find the integral J of $(ax^2 + bx + c)^{-1}$, with $a \neq 0$, distinguishing between the cases (i) $b^2 > 4ac$, (ii) $b^2 < 4ac$ and (iii) $b^2 = 4ac$.

In each case, we first 'complete the square' in the denominator, i.e. write it in such a form that x appears *only* in a term that is the square of a linear function of x. We then examine the overall sign of the terms that do not contain x; this determines the form of the integral. In case (iii) there is no such term. We write $b^2 - 4ac$ as $\Delta^2 > 0$, or $4ac - b^2$ as $\Delta'^2 > 0$, as needed.

(i) For $\Delta^2 = b^2 - 4ac > 0$,

$$J = \int \frac{dx}{a\left[\left(x + \frac{b}{2a}\right)^2 - \left(\frac{b^2}{4a^2} - \frac{c}{a}\right)\right]}$$

$$= \frac{1}{a} \int \frac{dx}{\left(x + \frac{b}{2a}\right)^2 - \frac{\Delta^2}{4a^2}}$$

$$= \frac{1}{a} \frac{a}{\Delta} \ln \frac{x + \frac{b}{2a} - \frac{\Delta}{2a}}{x + \frac{b}{2a} + \frac{\Delta}{2a}} + k$$

$$= \frac{1}{\Delta} \ln \frac{2ax + b - \Delta}{2ax + b + \Delta} + k.$$

(ii) For $-\Delta'^2 = b^2 - 4ac < 0$,

$$J = \int \frac{dx}{a\left[\left(x + \frac{b}{2a}\right)^2 - \left(\frac{b^2}{4a^2} - \frac{c}{a}\right)\right]}$$

$$= \frac{1}{a} \int \frac{dx}{\left(x + \frac{b}{2a}\right)^2 + \frac{\Delta'^2}{4a^2}}$$

$$= \frac{1}{a} \frac{2a}{\Delta'} \tan^{-1} \left(\frac{x + \frac{b}{2a}}{\frac{\Delta'}{2a}}\right) + k$$

$$= \frac{2}{\Delta'} \tan^{-1} \left(\frac{2ax + b}{\Delta'}\right) + k.$$

(iii) For $b^2 - 4ac = 0$,

$$J = \int \frac{dx}{ax^2 + bx + \frac{b^2}{4a}}$$

$$= \frac{1}{a} \int \frac{dx}{\left(x + \frac{b}{2a}\right)^2}$$

$$= \frac{-1}{a\left(x + \frac{b}{2a}\right)} + k$$

$$= -\frac{2}{2ax + b} + k.$$

4.7 Find the derivative of $f(x) = (1 + \sin x)/\cos x$ and hence determine the indefinite integral J of $\sec x$.

We differentiate $f(x)$ as a quotient, i.e. using $d(u/v)/dx = (vu' - uv')/v^2$, and obtain

$$f(x) = \frac{1 + \sin x}{\cos x},$$

$$f'(x) = \frac{\cos x (\cos x) - (1 + \sin x)(-\sin x)}{\cos^2 x}$$

$$= \frac{1 + \sin x}{\cos^2 x}$$

$$= \frac{f(x)}{\cos x}.$$

Thus, since $\sec x = f'(x)/f(x)$, it follows that

$$\int \sec x \, dx = \ln[f(x)] + c = \ln\left(\frac{1 + \sin x}{\cos x}\right) + c = \ln(\sec x + \tan x) + c.$$

4.9 By making the substitution $x = a \cos^2 \theta + b \sin^2 \theta$, evaluate the definite integrals J between limits a and b $(>a)$ of the following functions:

(a) $[(x - a)(b - x)]^{-1/2}$;
(b) $[(x - a)(b - x)]^{1/2}$;
(c) $[(x - a)/(b - x)]^{1/2}$.

Wherever the substitution $x = a \cos^2 \theta + b \sin^2 \theta$ is made, the terms in parentheses take the following forms:

$$x - a \to a \cos^2 \theta + b \sin^2 \theta - a = -a \sin^2 \theta + b \sin^2 \theta = (b - a) \sin^2 \theta,$$

$$b - x \to b - a \cos^2 \theta - b \sin^2 \theta = -a \cos^2 \theta + b \cos^2 \theta = (b - a) \cos^2 \theta,$$

and dx will be given by

$$dx = [2a \cos \theta (-\sin \theta) + 2b \sin \theta (\cos \theta)] \, d\theta = 2(b - a) \cos \theta \sin \theta \, d\theta.$$

The limits a and b will be replaced by 0 and $\pi/2$, respectively. We also note that the average value of the square of a sinusoid over any whole number of quarter cycles of its argument is one-half.

(a) $J_a = \displaystyle\int_a^b \frac{dx}{[(x - a)(b - x)]^{1/2}}$

$= \displaystyle\int_0^{\pi/2} \frac{2(b - a) \cos \theta \sin \theta}{[(b - a) \sin^2 \theta \, (b - a) \cos^2 \theta]^{1/2}} \, d\theta$

$= \displaystyle\int_0^{\pi/2} 2 \, d\theta = \pi.$

(b) $J_b = \displaystyle\int_a^b [(x-a)(b-x)]^{1/2}\,dx$

$= \displaystyle\int_0^{\pi/2} 2(b-a)^2 \cos^2\theta \sin^2\theta\,d\theta$

$= \dfrac{1}{2}(b-a)^2 \displaystyle\int_0^{\pi/2} \sin^2 2\theta\,d\theta$

$= \dfrac{1}{2}(b-a)^2 \dfrac{1}{2}\dfrac{\pi}{2} = \dfrac{\pi(b-a)^2}{8}.$

(c) $J_c = \displaystyle\int_a^b \sqrt{\dfrac{x-a}{b-x}}\,dx$

$= \displaystyle\int_0^{\pi/2} \sqrt{\dfrac{(b-a)\sin^2\theta}{(b-a)\cos^2\theta}} \times 2(b-a)\cos\theta\sin\theta\,d\theta$

$= \displaystyle\int_0^{\pi/2} 2(b-a)\sin^2\theta\,d\theta$

$= \dfrac{\pi(b-a)}{2}.$

4.11 Use integration by parts to evaluate the following:

(a) $\displaystyle\int_0^y x^2 \sin x\,dx;$ (b) $\displaystyle\int_1^y x \ln x\,dx;$

(c) $\displaystyle\int_0^y \sin^{-1} x\,dx;$ (d) $\displaystyle\int_1^y \ln(a^2+x^2)/x^2\,dx.$

If u and v are functions of x, the general formula for integration by parts is

$$\int_a^b uv'\,dx = [uv]_a^b - \int_a^b u'v\,dx.$$

Any given integrand $w(x)$ has to be written as $w(x) = u(x)v'(x)$ with $v'(x)$ chosen so that (i) it can be integrated explicitly and (ii) it results in a u that has u' no more complicated than u itself. There are usually several possible choices, but the one that makes both u and v as simple as possible is normally the best.

(a) Here the obvious choice at the first stage is $u(x) = x^2$ and $v'(x) = \sin x$. For the second stage, $u = x$ and $v' = \cos x$ are equally clear assignments.

$$\int_0^y x^2 \sin x\,dx = \left[x^2(-\cos x)\right]_0^y - \int_0^y 2x(-\cos x)\,dx$$

$$= -y^2 \cos y + [2x \sin x]_0^y - \int_0^y 2\sin x\,dx$$

$$= -y^2 \cos y + 2y \sin y + [2\cos x]_0^y$$

$$= (2 - y^2)\cos y + 2y \sin y - 2.$$

(b) This integration is most straightforwardly carried out by taking $v'(x) = x$ and $u(x) = \ln x$ as follows:

$$\int_1^y x \ln x \, dx = \left[\frac{x^2}{2} \ln x \right]_1^y - \int_1^y \frac{1}{x} \frac{x^2}{2} dx$$

$$= \frac{y^2}{2} \ln y - \left[\frac{x^2}{4} \right]_1^y$$

$$= \frac{1}{2} y^2 \ln y + \frac{1}{4}(1 - y^2).$$

However, if you know that the integral of $\ln x$ is $x \ln x - x$, then the given integral can also be found by taking $v' = \ln x$ and $u = x$:

$$\int_1^y x \ln x \, dx = [x(x \ln x - x)]_1^y - \int_1^y 1 \times (x \ln x - x) \, dx$$

$$= y^2 \ln y - y^2 - 0 + 1 - \int_1^y x \ln x \, dx + \left[\frac{x^2}{2} \right]_1^y.$$

After the limits have been substituted, the equation can be rearranged as

$$2 \int_1^y x \ln x \, dx = y^2 \ln y - y^2 + 1 + \frac{y^2}{2} - \frac{1}{2},$$

$$\int_1^y x \ln x \, dx = \frac{1}{2} y^2 \ln y + \frac{1}{4}(1 - y^2).$$

(c) Here we do not know the integral of $\sin^{-1} x$ (that is the problem!) but we do know its derivative. Therefore consider the integrand as $1 \times \sin^{-1} x$, with $v'(x) = 1$ and $u(x) = \sin^{-1} x$.

$$\int_0^y \sin^{-1} x \, dx = \int_0^y 1 \, \sin^{-1} x \, dx$$

$$= \left[x \sin^{-1} x \right]_0^y - \int_0^y \frac{1}{\sqrt{1 - x^2}} x \, dx$$

$$= y \sin^{-1} y + \left[\sqrt{1 - x^2} \right]_0^y$$

$$= y \sin^{-1} y + \sqrt{1 - y^2} - 1.$$

(d) When the logarithm of a function of x appears as part of an integrand, it is normally helpful to remove its explicit appearance by making it the $u(x)$ part of an integration-by-parts formula. The reciprocal of the function, without any explicit logarithm, then appears in the resulting integral; this is usually easier to deal with. In this case we take $\ln(a^2 + x^2)$

as $u(x)$.

$$\int_1^y \frac{\ln(a^2 + x^2)}{x^2} dx = \left[-\frac{\ln(a^2 + x^2)}{x} \right]_1^y - \int_1^y \frac{2x}{a^2 + x^2} \left(-\frac{1}{x} \right) dx$$

$$= -\frac{\ln(a^2 + y^2)}{y} + \ln(a^2 + 1) + \frac{2}{a} \left[\tan^{-1} \left(\frac{x}{a} \right) \right]_1^y$$

$$= -\frac{\ln(a^2 + y^2)}{y} + \ln(a^2 + 1) + \frac{2}{a} \left[\tan^{-1} \left(\frac{y}{a} \right) - \tan^{-1} \left(\frac{1}{a} \right) \right].$$

4.13 The gamma function $\Gamma(n)$ is defined for all $n > -1$ by

$$\Gamma(n + 1) = \int_0^\infty x^n e^{-x} dx.$$

Find a recurrence relation connecting $\Gamma(n + 1)$ and $\Gamma(n)$.

(a) Deduce (i) the value of $\Gamma(n + 1)$ when n is a non-negative integer and (ii) the value of $\Gamma \left(\frac{7}{2} \right)$, given that $\Gamma \left(\frac{1}{2} \right) = \sqrt{\pi}$.

(b) Now, taking factorial m for *any* m to be defined by $m! = \Gamma(m + 1)$, evaluate $\left(-\frac{3}{2} \right)!$.

Integrating the defining equation by parts,

$$\Gamma(n + 1) = \int_0^\infty x^n e^{-x} dx = \left[-x^n e^{-x} \right]_0^\infty + \int_0^\infty n x^{n-1} e^{-x} dx$$

$$= 0 + n\Gamma(n), \quad \text{for } n > 0,$$

i.e. $\Gamma(n + 1) = n\Gamma(n)$.

(a)(i) Clearly $\Gamma(n + 1) = n(n - 1)(n - 2) \cdots 2\, 1\, \Gamma(1)$. But

$$\Gamma(1) = \int_0^\infty e^{-x} dx = 1.$$

Hence $\Gamma(n + 1) = n!$.

(a)(ii) Applying the recurrence relation derived above,

$$\Gamma \left(\frac{7}{2} \right) = \frac{5}{2} \frac{3}{2} \frac{1}{2} \Gamma \left(\frac{1}{2} \right) = \frac{15}{8} \sqrt{\pi}.$$

(b) With this general definition of a factorial, we have

$$\left(-\frac{3}{2} \right)! = \Gamma \left(-\frac{1}{2} \right) = \frac{1}{-1/2} \Gamma \left(\frac{1}{2} \right) = -2\sqrt{\pi}.$$

4.15 By integrating by parts twice, prove that I_n, as defined in the first equality below for positive integers n, has the value given in the second equality:

$$I_n = \int_0^{\pi/2} \sin n\theta \cos \theta \, d\theta = \frac{n - \sin(n\pi/2)}{n^2 - 1}.$$

Taking $\sin n\theta$ as u and $\cos \theta$ as v and noting that with this choice $u'' = -n^2 u$ and $v'' = -v$, we expect that after two integrations by parts we will recover (a multiple of) I_n.

$$I_n = \int_0^{\pi/2} \sin n\theta \cos \theta \, d\theta$$

$$= [\sin n\theta \sin \theta]_0^{\pi/2} - \int_0^{\pi/2} n \cos n\theta \sin \theta \, d\theta$$

$$= \sin \frac{n\pi}{2} - n \left\{ [-\cos n\theta \cos \theta]_0^{\pi/2} - \int_0^{\pi/2} (-n \sin n\theta)(-\cos \theta) \, d\theta \right\}$$

$$= \sin \frac{n\pi}{2} - n[-(-1) - n I_n].$$

Rearranging this gives

$$I_n(1 - n^2) = \sin \frac{n\pi}{2} - n,$$

and hence the stated result.

4.17 If J_r is the integral

$$\int_0^\infty x^r \exp(-x^2) \, dx,$$

show that

(a) $J_{2r+1} = (r!)/2,$
(b) $J_{2r} = 2^{-r}(2r - 1)(2r - 3) \cdots (5)(3)(1) J_0.$

(a) We first derive a recurrence relationship for J_{2r+1}. Since we cannot integrate $\exp(-x^2)$ explicitly but can integrate $-2x \exp(-x^2)$, we extract the factor $-2x$ from the rest of the integrand and treat what is left $(-\frac{1}{2}x^{2r}$ in this case) as $u(x)$. This is the operation that has been carried out in the second line of what follows.

$$J_{2r+1} = \int_0^\infty x^{2r+1} \exp(-x^2) \, dx$$

$$= \int_0^\infty -\frac{x^{2r}}{2}(-2x) \exp(-x^2) \, dx$$

$$= \left[-\frac{x^{2r}}{2} \exp(-x^2) \right]_0^\infty + \int_0^\infty \frac{2r x^{2r-1}}{2} \exp(-x^2) \, dx$$

$$= 0 + r J_{2r-1}.$$

Applying the relationship r times gives

$$J_{2r+1} = r(r-1) \cdots 1 J_1.$$

But

$$J_1 = \int_0^\infty x \exp(-x^2)\,dx = \left[-\frac{1}{2}\exp(-x^2) \right]_0^\infty = \frac{1}{2},$$

and so $J_{2r+1} = \frac{1}{2}r!$.

(b) Using the same method as in part (a) it can be shown that

$$J_{2r} = \frac{2r-1}{2}J_{2r-2}.$$

Hence,

$$J_{2r} = \frac{2r-1}{2}\frac{2r-3}{2}\cdots\frac{1}{2}J_0,$$

in agreement with the stated relationship.

4.19 By noting that for $0 \le \eta \le 1$, $\eta^{1/2} \ge \eta^{3/4} \ge \eta$, prove that

$$\frac{2}{3} \le \frac{1}{a^{5/2}} \int_0^a (a^2 - x^2)^{3/4}\,dx \le \frac{\pi}{4}.$$

We use the result that if $g(x) \le f(x) \le h(x)$ for *all* x in the range $a \le x \le b$, then $\int g(x)\,dx \le \int f(x)\,dx \le \int h(x)\,dx$, where all integrals are between the limits a and b.

Set $\eta = 1 - (x/a)^2$ in the stated inequalities and integrate the result from 0 to a, giving

$$\int_0^a \left(1 - \frac{x^2}{a^2}\right)^{1/2} dx \ge \int_0^a \left(1 - \frac{x^2}{a^2}\right)^{3/4} dx \ge \int_0^a \left(1 - \frac{x^2}{a^2}\right) dx.$$

Substituting $x = a\sin\theta$ and $dx = a\cos\theta\,d\theta$ in the first term and carrying out the elementary integration in the third term yields

$$\int_0^{\pi/2} a\cos^2\theta\,d\theta \ge \frac{1}{a^{3/2}}\int_0^a (a^2 - x^2)^{3/4}\,dx \ge \left[x - \frac{x^3}{3a^2} \right]_0^a,$$

$$\Rightarrow \quad a\frac{1}{2}\frac{\pi}{2} \ge \frac{1}{a^{3/2}}\int_0^a (a^2 - x^2)^{3/4}\,dx \ge \frac{2a}{3},$$

$$\Rightarrow \quad \frac{\pi}{4} \ge \frac{1}{a^{5/2}}\int_0^a (a^2 - x^2)^{3/4}\,dx \ge \frac{2}{3}.$$

4.21 A vase has curved sides that are generated by rotating the part of the curve $x = \frac{1}{2}a(e^{y/a} + e^{-y/a})$ that lies between $y = 0$ and $y = ha$ around the y-axis. Show that the area of the curved surface is $\pi a^2 [\frac{1}{4}(e^{2h} - e^{-2h}) + h]$.

Since the height of the vase is given in terms of two values of y, we use the formula for its surface area in the form

$$A = \int_0^{ha} 2\pi x \sqrt{1 + \left(\frac{dx}{dy}\right)^2} \, dy$$

$$= \pi a \int_0^{ha} \left(e^{y/a} + e^{-y/a}\right) \left[1 + \frac{a^2}{4a^2} \left(e^{y/a} - e^{-y/a}\right)^2\right]^{1/2} dy$$

$$= \pi a \int_0^{ha} \left(e^{y/a} + e^{-y/a}\right) \left[\frac{1}{4} \left(4 + e^{2y/a} - 2 + e^{-2y/a}\right)\right]^{1/2} dy$$

$$= \pi a \int_0^{ha} \left(e^{y/a} + e^{-y/a}\right) \frac{1}{2} \left(e^{y/a} + e^{-y/a}\right) dy$$

$$= \frac{\pi a}{2} \int_0^{ha} \left(e^{2y/a} + 2 + e^{-2y/a}\right) dy$$

$$= \frac{\pi a}{2} \left(\left[\frac{ae^{2y/a}}{2}\right]_0^{ha} + 2ha + \left[\frac{-ae^{-2y/a}}{2}\right]_0^{ha}\right)$$

$$= \frac{\pi a^2}{2} \left(\frac{e^{2h}}{2} - \frac{1}{2} + 2h - \frac{e^{-2h}}{2} + \frac{1}{2}\right)$$

$$= \pi a^2 \left[\frac{1}{4} \left(e^{2h} - e^{-2h}\right) + h\right].$$

Note: The algebra would be less lengthy to write out if hyperbolic sines and cosines were used, but they are not introduced until Chapter 5.

4.23 By noting that $\sinh x < \frac{1}{2}e^x < \cosh x$, and that $1 + z^2 < (1+z)^2$ for $z > 0$, show that, for $x > 0$, the length L of the curve $y = \frac{1}{2}e^x$ measured from the origin satisfies the inequalities $\sinh x < L < x + \sinh x$.

With $y = y' = \frac{1}{2}e^x$ and the element of curve length ds given by $ds = (1 + y'^2)^{1/2} dx$, the total length of the curve measured from the origin is

$$L = \int_0^x ds = \int_0^x \left(1 + \tfrac{1}{4}e^{2x}\right)^{1/2} dx.$$

But, since all quantities are positive for $x \geq 0$,

$$\sinh x < \quad \tfrac{1}{2}e^x \quad < \cosh x,$$

$$\Rightarrow \quad \sinh^2 x < \quad \tfrac{1}{4}e^{2x} \quad < \cosh^2 x,$$

$$\cosh^2 x = 1 + \sinh^2 x < \quad 1 + \tfrac{1}{4}e^{2x} \quad < 1 + \cosh^2 x < (1 + \cosh x)^2,$$

$$\Rightarrow \quad \cosh x < \left(1 + \tfrac{1}{4}e^{2x}\right)^{1/2} < 1 + \cosh x.$$

It then follows, from integrating each term in the double inequality, that

$$\int_0^x \cosh x \, dx \; < \; L \; < \; \int_0^x (1 + \cosh x) \, dx,$$

$$\Rightarrow \qquad \sinh x \; < \; L \; < \; x + \sinh x,$$

as stated in the question.

5 Complex numbers and hyperbolic functions

5.1 Express the following as single complex numbers.

    (a) $(3 + 2i) + (-1 + i) - (5 + 2i)$,   (b) $(3 + 2i)(4 - 3i)$,   (c) $(3 + 2i)/(4 - 3i)$.

    (a) $(3 + 2i) + (-1 + i) - (5 + 2i) = (3 - 1 - 5) + i(2 + 1 - 2) = -3 + i$.
    (b) $(3 + 2i)(4 - 3i) = 12 + 8i - 9i - 6i^2 = 18 - i$.
    (c) $\dfrac{3 + 2i}{4 - 3i} = \dfrac{(3 + 2i)(4 + 3i)}{(4 - 3i)(4 + 3i)} = \dfrac{12 + 8i + 9i + 6i^2}{16 + 9} = \dfrac{6 + 17i}{25}$.

5.3 Two complex numbers z and w are given by $z = 3 + 4i$ and $w = 2 - i$. On an Argand diagram, plot

    (a) $z + w$,   (b) $w - z$,   (c) wz,   (d) z/w,
    (e) $z^*w + w^*z$,   (f) w^2,   (g) $\ln z$,   (h) $(1 + z + w)^{1/2}$.

With $z = 3 + 4i$, $w = 2 - i$ and, where needed, $i^2 = -1$:

    (a) $z + w = 3 + 4i + 2 - i = 5 + 3i$;
    (b) $w - z = 2 - i - 3 - 4i = -1 - 5i$;
    (c) $wz = (2 - i)(3 + 4i) = 6 - 3i + 8i - 4i^2 = 10 + 5i$;
    (d) $\dfrac{z}{w} = \dfrac{3 + 4i}{2 - i} = \dfrac{3 + 4i}{2 - i}\dfrac{2 + i}{2 + i} = \dfrac{6 + 8i + 3i + 4i^2}{4 - 2i + 2i - i^2} = \dfrac{2 + 11i}{5}$;
    (e) $z * w + w * z = (3 - 4i)(2 - i) + (2 + i)(3 + 4i) = (2 - 11i) + (2 + 11i) = 4$;
    (f) $w^2 = (2 - i)(2 - i) = 4 - 4i + i^2 = 3 - 4i$;
    (g) $\ln(3 + 4i) = \ln(3^2 + 4^2)^{1/2} + i \tan^{-1}\left(\frac{4}{3}\right) = \ln 5 + i\left[\tan^{-1}\left(\frac{4}{3}\right) + 2n\pi\right]$;
    (h) $(1 + z + w)^{1/2} = (6 + 3i)^{1/2}$

$$(6 + 3i)^{1/2} = \left\{\sqrt{45}\, \exp\left[i \tan\left(\tfrac{3}{6}\right)\right]\right\}^{1/2}$$
$$= \pm(45)^{1/4} \exp\left[i\tfrac{1}{2} \tan^{-1}\left(\tfrac{1}{2}\right)\right]$$
$$= \pm 2.590\,(\cos 0.2318 + i \sin 0.2318)$$
$$= \pm(2.521 + 0.595i).$$

These results are plotted in Figure 5.1. The answer to part (g) is multivalued and only five of the infinite number of possibilities are shown.

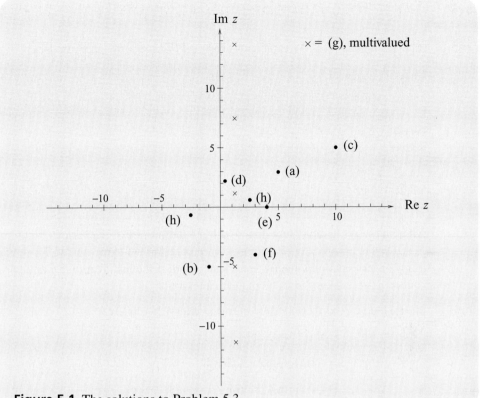

Figure 5.1 The solutions to Problem 5.3.

5.5 By writing $\pi/12 = (\pi/3) - (\pi/4)$ and considering $e^{i\pi/12}$, evaluate $\cot(\pi/12)$.

As we are expressing $\pi/12$ as the difference between two (familiar) angles, for which we know explicit formulae for their sines and cosines, namely

$$\sin\frac{\pi}{3} = \frac{\sqrt{3}}{2}, \quad \cos\frac{\pi}{3} = \frac{1}{2}, \quad \sin\frac{\pi}{4} = \cos\frac{\pi}{4} = \frac{1}{\sqrt{2}},$$

we will need the formulae for $\cos(A - B)$ and $\sin(A - B)$. They are given by

$$\cos(A - B) = \cos A \cos B + \sin A \sin B$$
$$\text{and} \quad \sin(A - B) = \sin A \cos B - \cos A \sin B.$$

Applying these with $A = \pi/3$ and $B = \pi/4$,

$$\exp\left(i\frac{\pi}{12}\right) = \exp\left[i\left(\frac{\pi}{3} - \frac{\pi}{4}\right)\right],$$
$$\cos\frac{\pi}{12} + i\sin\frac{\pi}{12} = \cos\left(\frac{\pi}{3} - \frac{\pi}{4}\right) + i\sin\left(\frac{\pi}{3} - \frac{\pi}{4}\right)$$

$$= \cos\frac{\pi}{3}\cos\frac{\pi}{4} + \sin\frac{\pi}{3}\sin\frac{\pi}{4} + i\left(\sin\frac{\pi}{3}\cos\frac{\pi}{4} - \cos\frac{\pi}{3}\sin\frac{\pi}{4}\right)$$

$$= \left(\frac{1}{2}\frac{1}{\sqrt{2}} + \frac{\sqrt{3}}{2}\frac{1}{\sqrt{2}}\right) + i\left(\frac{\sqrt{3}}{2}\frac{1}{\sqrt{2}} - \frac{1}{2}\frac{1}{\sqrt{2}}\right).$$

Thus

$$\cot\frac{\pi}{12} = \frac{\cos(\pi/12)}{\sin(\pi/12)} = \frac{1+\sqrt{3}}{\sqrt{3}-1} = 2 + \sqrt{3}.$$

5.7 Evaluate

(a) $\mathrm{Re}(\exp 2iz)$,   (b) $\mathrm{Im}(\cosh^2 z)$,   (c) $(-1+\sqrt{3}i)^{1/2}$,   (d) $|\exp(i^{1/2})|$,
(e) $\exp(i^3)$,   (f) $\mathrm{Im}(2^{i+3})$,   (g) i^i,   (h) $\ln[(\sqrt{3}+i)^3]$.

All of these evaluations rely directly on the definitions of the various functions involved as applied to complex numbers; these should be known to the reader. There are too many to give every one individually at each step and, if the justification for any particular step is unclear, reference should be made to a textbook.

(a) $\mathrm{Re}(\exp 2iz) = \mathrm{Re}[\exp(2ix - 2y)] = \exp(-2y)\cos 2x.$

(b) $\mathrm{Im}(\cosh^2 z) = \mathrm{Im}\left[\frac{1}{2}(\cosh 2z + 1)\right]$

$$= \frac{1}{2}\,\mathrm{Im}[\cosh(2x + 2iy)]$$

$$= \frac{1}{2}\,\mathrm{Im}(\cosh 2x \cosh 2iy + \sinh 2x \sinh 2iy)$$

$$= \frac{1}{2}\,\mathrm{Im}(\cosh 2x \cos 2y + i \sinh 2x \sin 2y)$$

$$= \frac{1}{2}\,\sinh 2x \sin 2y.$$

(c) $(-1+\sqrt{3}i)^{1/2} = \left[(-1)^2 + (\sqrt{3})^2\right]^{1/4}\exp\left[i\frac{1}{2}(\tan^{-1}(\frac{\sqrt{3}}{-1}) + 2n\pi)\right]$

$$= \sqrt{2}\exp\left[i\frac{1}{2}(\frac{2}{3}\pi + 2n\pi)\right]$$

$$= \sqrt{2}\exp\left(\frac{\pi i}{3}\right) \quad \text{or} \quad \sqrt{2}\exp\left(\frac{4\pi i}{3}\right).$$

(d) $\left|\exp(i^{1/2})\right| = \left|\exp[(e^{i\pi/2})^{1/2}]\right|$

$$= \left|\exp[e^{i\pi/4 + in\pi}]\right|$$

$$= \left|\exp[\cos(n + \tfrac{1}{4})\pi + i\sin(n + \tfrac{1}{4})\pi]\right|$$

$$= \exp[\cos(n + \tfrac{1}{4})\pi]$$

$$= \exp\left(\tfrac{1}{\sqrt{2}}\right) \quad \text{or} \quad \exp\left(-\tfrac{1}{\sqrt{2}}\right).$$

(e) $\exp(i^3) = \exp[e^{3(i\pi/2)}] = \exp(\cos\frac{3\pi}{2} + i\sin\frac{3\pi}{2})$

$$= \exp(0 - i) = \cos(-1) + i\sin(-1) = 0.540 - 0.841\,i.$$

(f) $\mathrm{Im}(2^{i+3}) = \mathrm{Im}(8 \times 2^i) = 8\,\mathrm{Im}(2^i) = 8\,\mathrm{Im}(e^{i\ln 2}) = 8\,\sin(\ln 2) = 5.11.$

(g) $i^i = \left[\exp i(\tfrac{1}{2}\pi + 2n\pi) \right]^i = \left[\exp i^2(\tfrac{1}{2}\pi + 2n\pi) \right] = \exp[-(2n + \tfrac{1}{2})\pi\,]$.

(h) $\ln\left[(\sqrt{3} + i)^3\right] = 3\ln(\sqrt{3} + i)$

$$= 3\left(\ln 2 + i\tan^{-1}\tfrac{1}{\sqrt{3}}\right)$$

$$= \ln 8 + 3i(\tfrac{\pi}{6} + 2n\pi)$$

$$= \ln 8 + i(6n + \tfrac{1}{2})\pi.$$

5.9 The two sets of points $z = a$, $z = b$, $z = c$, and $z = A$, $z = B$, $z = C$ are the corners of two similar triangles in the Argand diagram. Express in terms of $a, b, \ldots, C$

(a) the equalities of corresponding angles and
(b) the constant ratio of corresponding sides,

in the two triangles.
    By noting that any complex quantity can be expressed as

$$z = |z|\exp(i\arg z),$$

deduce that

$$a(B - C) + b(C - A) + c(A - B) = 0.$$

(a) The equalities of corresponding angles is expressed by

$$\arg(A - B) = \arg(a - b), \quad \arg(B - C) = \arg(b - c), \quad \arg(C - A) = \arg(c - a).$$

(b) The constant ratio of the sides is represented by the equation

$$\frac{|A - B|}{|a - b|} = \frac{|B - C|}{|b - c|} = \frac{|C - A|}{|c - a|} = \lambda.$$

Now consider the expression

$$z = a(B - C) + b(C - A) + c(A - B)$$

$$= a|B - C|\exp[i\arg(B - C)] + b|C - A|\exp[i\arg(C - A)]$$

$$+ c|A - B|\exp[i\arg(A - B)]$$

Now, using the results of (a) and (b) to replace the properties of triangle ABC with those of triangle abc, we have

$$z = \lambda a|b - c|\exp[i\arg(b - c)] + \lambda b|c - a|\exp[i\arg(c - a)]$$

$$+ \lambda c|a - b|\exp[i\arg(a - b)]$$

$$= \lambda[a(b - c) + b(c - a) + c(a - b)]$$

$$= \lambda[ab - ac + bc - ba + ca - cb)] = 0.$$

In the final equality we have used the fact that multiplication of complex numbers is commutative.

5.11 Sketch the parts of the Argand diagram in which

(a) $\operatorname{Re} z^2 < 0$, $|z^{1/2}| \leq 2$;
(b) $0 \leq \arg z^* \leq \pi/2$;
(c) $|\exp z^3| \to 0$ as $|z| \to \infty$.

What is the area of the region in which all three sets of conditions are satisfied?

Since we will need to study the signs of the real parts of certain powers of z, it will be convenient to consider z as $r\,e^{i\theta}$ with $0 \leq \theta \leq 2\pi$.

Condition (a) contains two specifications. Firstly, for the real part of z^2 to be negative, its argument must be greater than $\pi/2$ but less than $3\pi/2$. The argument of z itself, which is half that of z^2 (mod 2π), must therefore lie in one of the two ranges $\pi/4 < \arg z < 3\pi/4$ and $5\pi/4 < \arg z < 7\pi/4$. Secondly, since the modulus of any complex number is real and positive, $|z^{1/2}| \leq 2$ is equivalent to $|z| \leq 4$.

Since $\arg z^* = -\arg z$, condition (b) requires $\arg z$ to lie in the range $3\pi/2 \leq \theta \leq 2\pi$, i.e z to lie in the fourth quadrant.

Condition (c) will only be satisfied if the real part of z^3 is negative. This requires

$$(4n+1)\frac{\pi}{2} < 3\theta < (4n+3)\frac{\pi}{2}, \qquad n = 0, 1, 2.$$

The allowed regions for θ are thus alternate wedges of angular size $\pi/3$ with an allowed region starting at $\theta = \pi/6$. The allowed region overlapping those specified by conditions (a) and (b) is the wedge $3\pi/2 \leq \theta \leq 11\pi/6$.

All three conditions are satisfied in the region $3\pi/2 \leq \theta \leq 7\pi/4$, $|z| \leq 4$; see Figure 5.2. This wedge has an area given by

$$\frac{1}{2}r^2\theta = \frac{1}{2}16\left(\frac{7\pi}{4} - \frac{3\pi}{2}\right) = 2\pi.$$

5.13 Prove that $x^{2m+1} - a^{2m+1}$, where m is an integer ≥ 1, can be written as

$$x^{2m+1} - a^{2m+1} = (x-a)\prod_{r=1}^{m}\left[x^2 - 2ax\cos\left(\frac{2\pi r}{2m+1}\right) + a^2\right].$$

For the sake of brevity, we will denote $x^{2m+1} - a^{2m+1}$ by $f(x)$ and the $(2m+1)$th root of unity, $\exp[\,2\pi i/(2m+1)\,]$, by Ω.

Now consider the roots of the equation $f(x) = 0$. The $2m+1$ quantities of the form $x = a\Omega^r$ with $r = 0, 1, 2, \ldots, 2m$ are all solutions of this equation and, since it is a polynomial equation of order $2m+1$, they represent all of its roots. We can therefore

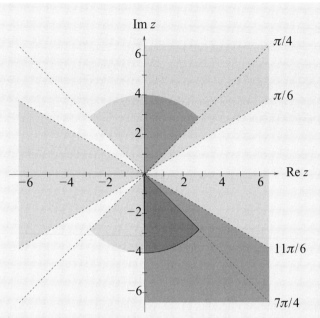

Figure 5.2 The defined region of the Argand diagram in Problem 5.11. Regions in which only one condition is satisfied are lightly shaded; those that satisfy two conditions are more heavily shaded; and the region satisfying all three conditions is most heavily shaded and outlined.

reconstruct the polynomial $f(x)$ (which has unity as the coefficient of its highest power) as the product of factors of the form $(x - a\Omega^r)$:

$$f(x) = (x - a)(x - a\Omega)\cdots(x - a\Omega^m)(x - a\Omega^{m+1})\cdots(x - a\Omega^{2m}).$$

Now combine $(x - a\Omega^r)$ with $(x - a\Omega^{2m+1-r})$:

$$f(x) = (x - a)\prod_{r=1}^{m}(x - a\Omega^r)(x - a\Omega^{2m+1-r})$$

$$= (x - a)\prod_{r=1}^{m}[x^2 - ax(\Omega^r + \Omega^{2m+1-r}) + a^2\Omega^{2m+1}]$$

$$= (x - a)\prod_{r=1}^{m}[x^2 - ax(\Omega^r + \Omega^{-r}) + a^2], \quad \text{since } \Omega^{2m+1} = 1,$$

$$= (x - a)\prod_{r=1}^{m}\left[x^2 - 2ax\cos\left(\frac{2\pi r}{2m + 1}\right) + a^2\right].$$

This is the form given in the question.

5.15 Solve the equation

$$z^7 - 4z^6 + 6z^5 - 6z^4 + 6z^3 - 12z^2 + 8z + 4 = 0,$$

(a) by examining the effect of setting $z^3 = 2$ and then
(b) by factorising and using the binomial expansion of $(z + a)^4$.

Plot the seven roots of the equation on an Argand diagram, exemplifying that complex roots of a polynomial equation always occur in conjugate pairs if the polynomial has real coefficients.

(a) Setting $z^3 = 2$ in $f(z)$ so as to leave no higher powers of z than its square, e.g. writing z^7 as $(z^3)^2 z = 4z$, gives

$$4z - 16 + 12z^2 - 12z + 12 - 12z^2 + 8z + 4 = 0,$$

which is satisfied identically. Thus $z^3 - 2$ is a factor of $f(z)$.
    (b) Writing $f(z)$ as

$$f(z) = (z^3 - 2)(az^4 + bz^3 + cz^2 + dz + e) = 0$$

and equating the coefficients of the various powers of z gives $a = 1$, $b = -4$, $c = 6$, $d - 2a = -6$, $e - 2b = 6$, $-2c = -12$, $-2d = 8$ and $-2e = 4$. These imply (consistently) that $f(z)$ can be written as

$$f(z) = (z^3 - 2)(z^4 - 4z^3 + 6z^2 - 4z - 2).$$

We now note that the first four terms in the second set of parentheses are the same as the corresponding terms in the expansion of $(z - 1)^4$; only the constant term needs correction. Thus, we may write the original equation as

$$0 = f(z) = (z^3 - 2)[(z - 1)^4 - 3],$$

with solutions $\quad z = 2^{1/3} e^{2n\pi i/3} \quad n = 0,\ 1,\ 2 \quad$ or

$$z - 1 = 3^{1/4} e^{2n\pi i/4} \quad n = 0,\ 1,\ 2,\ 3.$$

The seven roots, which can now be directly plotted on an Argand diagram, are therefore

$$2^{1/3}, \quad 2^{1/3}\left(\frac{-1 \pm i\sqrt{3}}{2}\right), \quad 1 \pm 3^{1/4}, \quad 1 \pm 3^{1/4} i.$$

As is to be expected, each root that has a non-zero imaginary part occurs as one of a complex conjugate pair.

5.17 The binomial expansion of $(1 + x)^n$ can be written for a positive integer n as

$$(1 + x)^n = \sum_{r=0}^{n} {}^nC_r x^r,$$

where ${}^nC_r = n!/[r!(n - r)!]$.

(a) Use de Moivre's theorem to show that the sum

$$S_1(n) = {}^nC_0 - {}^nC_2 + {}^nC_4 - \cdots + (-1)^m \, {}^nC_{2m}, \qquad n-1 \le 2m \le n,$$

has the value $2^{n/2} \cos(n\pi/4)$.

(b) Derive a similar result for the sum

$$S_2(n) = {}^nC_1 - {}^nC_3 + {}^nC_5 - \cdots + (-1)^m \, {}^nC_{2m+1}, \qquad n-1 \le 2m+1 \le n,$$

and verify it for the cases $n = 6, 7$ and 8.

(a) Since we seek the sum of binomial coefficients that contain either all even or all odd indices, we need to choose a value for x such that x^r has different characteristics depending upon whether r is even or odd. The quantity i has just such a property, being purely real when r is even and purely imaginary when r is odd. We therefore take $x = i$, write $1 + i$ as $\sqrt{2}e^{i\pi/4}$ and apply de Moivre's theorem:

$$\begin{aligned}
\left(\sqrt{2}e^{i\pi/4}\right)^n &= (1+i)^n \\
&= {}^nC_0 + i \, {}^nC_1 + i^2 \, {}^nC_2 + \cdots \\
&= \left({}^nC_0 - {}^nC_2 + {}^nC_4 - \cdots\right) + i\left({}^nC_1 - {}^nC_3 + {}^nC_5 - \cdots\right).
\end{aligned}$$

Thus $S_1(n) = \left({}^nC_0 - {}^nC_2 + {}^nC_4 - \cdots + (-1)^m \, {}^nC_{2m}\right)$, where $n - 1 \le 2m \le n$, has a value equal to that of the real part of $(\sqrt{2}e^{i\pi/4})^n$. This is the real part of $2^{n/2}e^{in\pi/4}$, which, by de Moivre's theorem, is $2^{n/2} \cos(n\pi/4)$.

(b) The corresponding result for $S_2(n)$ is that it is equal to the imaginary part of $2^{n/2}e^{in\pi/4}$, which is $2^{n/2} \sin(n\pi/4)$.

We now verify this result for $n = 6, 7$ and 8 by direct calculation:

$$S_2(6) = {}^6C_1 - {}^6C_3 + {}^6C_5 = 6 - 20 + 6 = -8 = 2^3 \sin\frac{6\pi}{4},$$

$$\begin{aligned}
S_2(7) &= {}^7C_1 - {}^7C_3 + {}^7C_5 - {}^7C_7 \\
&= 7 - 35 + 21 - 1 = -8 = 2^{7/2} \sin\frac{7\pi}{4},
\end{aligned}$$

$$S_2(8) = {}^8C_1 - {}^8C_3 + {}^8C_5 - {}^8C_7 = 8 - 56 + 56 - 8 = 0 = 2^4 \sin\frac{8\pi}{4}.$$

5.19 Use de Moivre's theorem with $n = 4$ to prove that

$$\cos 4\theta = 8\cos^4\theta - 8\cos^2\theta + 1,$$

and deduce that

$$\cos\frac{\pi}{8} = \left(\frac{2+\sqrt{2}}{4}\right)^{1/2}.$$

From de Moivre's theorem, $e^{i4\theta} = \cos 4\theta + i \sin 4\theta$. But, by the binomial theorem, we also have that

$$e^{i4\theta} = (\cos \theta + i \sin \theta)^4$$
$$= \cos^4 \theta + 4i \cos^3 \theta \sin \theta - 6 \cos^2 \theta \sin^2 \theta - 4i \cos \theta \sin^3 \theta + \sin^4 \theta.$$

Equating the real parts of the two equal expressions and writing $\sin^2 \theta$ as $1 - \cos^2 \theta$,

$$\cos 4\theta = \cos^4 \theta - 6 \cos^2 \theta (1 - \cos^2 \theta) + (1 - \cos^2 \theta)^2$$
$$= 8 \cos^4 \theta - 8 \cos^2 \theta + 1.$$

Now set $\theta = \pi/8$ in this result and write $\cos(\pi/8)$ as c:

$$0 = \cos \frac{4\pi}{8} = 8c^4 - 8c^2 + 1.$$

Hence, as this is a quadratic equation in c^2,

$$c^2 = \frac{4 \pm \sqrt{16 - 8}}{8} \quad \text{and} \quad c = \cos \frac{\pi}{8} = \pm \left(\frac{2 \pm \sqrt{2}}{4} \right)^{1/2}.$$

Since $0 < \pi/8 < \pi/2$, c must be positive. Further, as $\pi/8 < \pi/4$ and $\cos(\pi/4) = 1/\sqrt{2}$, c must be greater then $1/\sqrt{2}$. It is clear that the positive square roots are the appropriate ones in both cases.

5.21 Use de Moivre's theorem to prove that

$$\tan 5\theta = \frac{t^5 - 10t^3 + 5t}{5t^4 - 10t^2 + 1},$$

where $t = \tan \theta$. Deduce the values of $\tan(n\pi/10)$ for $n = 1, 2, 3$ and 4.

Using the binomial theorem and de Moivre's theorem to expand $(e^{i\theta})^5$ in two different ways, we have, from equating the real and imaginary parts of the two results, that

$$\cos 5\theta + i \sin 5\theta = \cos^5 \theta + i5 \cos^4 \theta \sin \theta - 10 \cos^3 \theta \sin^2 \theta - i10 \cos^2 \theta \sin^3 \theta$$
$$+ 5 \cos \theta \sin^4 \theta + i \sin^5 \theta,$$
$$\cos 5\theta = \cos^5 \theta - 10 \cos^3 \theta (1 - \cos^2 \theta) + 5 \cos \theta (1 - 2 \cos^2 \theta + \cos^4 \theta)$$
$$= 16 \cos^5 \theta - 20 \cos^3 \theta + 5 \cos \theta,$$
$$\sin 5\theta = 5(1 - 2 \sin^2 \theta + \sin^4 \theta) \sin \theta - 10(1 - \sin^2 \theta) \sin^3 \theta + \sin^5 \theta$$
$$= 16 \sin^5 \theta - 20 \sin^3 \theta + 5 \sin \theta.$$

Now, writing $\cos\theta$ as c, $\sin\theta$ as s and $\tan\theta$ as t, and further recalling that $c^{-2} = 1 + t^2$, we have

$$
\begin{aligned}
\tan 5\theta &= \frac{16s^5 - 20s^3 + 5s}{16c^5 - 20c^3 + 5c} \\
&= \frac{16t^5 - 20t^3 c^{-2} + 5tc^{-4}}{16 - 20c^{-2} + 5c^{-4}} \\
&= \frac{16t^5 - 20t^3(1+t^2) + 5t(1 + 2t^2 + t^4)}{16 - 20(1+t^2) + 5(1 + 2t^2 + t^4)} \\
&= \frac{t^5 - 10t^3 + 5t}{5t^4 - 10t^2 + 1}.
\end{aligned}
$$

When θ is equal to $\dfrac{\pi}{10}$ or $\dfrac{3\pi}{10}$, $\tan 5\theta = \infty$, implying that

$$
5t^4 - 10t^2 + 1 = 0 \quad\Rightarrow\quad t^2 = \frac{5 \pm \sqrt{25 - 5}}{5} \quad\Rightarrow\quad t = \pm\left(\frac{5 \pm \sqrt{20}}{5}\right)^{1/2}.
$$

As both angles lie in the first quadrant the overall sign must be taken as positive in both cases, and it is clear that the positive square root in the numerator corresponds to $\theta = 3\pi/10$.

When θ is equal to $\dfrac{2\pi}{10}$ or $\dfrac{4\pi}{10}$, $\tan 5\theta = 0$, implying that

$$
t^5 - 10t^3 + 5t = 0 \quad\Rightarrow\quad t^2 = 5 \pm \sqrt{25 - 5} \quad\Rightarrow\quad t = \pm\left(5 \pm \sqrt{20}\right)^{1/2}.
$$

Again, as both angles lie in the first quadrant the overall sign must be taken as positive; it is also clear that the positive square root in the parentheses corresponds to $\theta = 4\pi/10$.

5.23 Determine the conditions under which the equation

$$
a\cosh x + b\sinh x = c, \qquad c > 0,
$$

has zero, one or two real solutions for x. What is the solution if $a^2 = c^2 + b^2$?

We start by recalling that $\cosh x = \frac{1}{2}(e^x + e^{-x})$ and $\sinh x = \frac{1}{2}(e^x - e^{-x})$, and then rewrite the equation as a quadratic equation in e^x:

$$
\begin{aligned}
a\cosh x + b\sinh x - c &= 0, \\
(a + b)e^x - 2c + (a - b)e^{-x} &= 0, \\
(a + b)e^{2x} - 2ce^x + (a - b) &= 0.
\end{aligned}
$$

Hence,

$$
e^x = \frac{c \pm \sqrt{c^2 - (a^2 - b^2)}}{a + b}.
$$

For x to be real, e^x must be real and ≥ 0. Since $c > 0$, this implies that $a + b > 0$ and $c^2 + b^2 \geq a^2$. Provided these two conditions are satisfied, there are two roots if $c^2 + b^2 - a^2 < c^2$, i.e. if $b^2 < a^2$, but only one root if $c^2 + b^2 - a^2 > c^2$, i.e. if $b^2 > a^2$. If $c^2 + b^2 = a^2$ then the double root is given by

$$e^x = \frac{c}{a + b},$$

$$e^{2x} = \frac{c^2}{(a + b)^2} = \frac{a^2 - b^2}{(a + b)^2} = \frac{a - b}{a + b},$$

$$x = \frac{1}{2} \ln \frac{a - b}{a + b}.$$

5.25 Express $\sinh^4 x$ in terms of hyperbolic cosines of multiples of x, and hence find the real solutions of

$$2 \cosh 4x - 8 \cosh 2x + 5 = 0.$$

In order to connect $\sinh^4 x$ to hyperbolic functions of other multiples of x, we need to express it in terms of powers of $e^{\pm x}$ and then to group the terms so as to make up those hyperbolic functions. Starting from

$$\sinh x = \tfrac{1}{2}(e^x - e^{-x}),$$

we have from the binomial theorem that

$$\sinh^4 x = \tfrac{1}{16}\left(e^{4x} - 4e^{2x} + 6 - 4e^{-2x} + e^{-4x}\right).$$

Terms containing related exponents nx and $-nx$ can now be grouped together and expressed as a linear sum of $\cosh nx$ and $\sinh nx$; here, because of the symmetry properties of the binomial coefficients, only the $\cosh nx$ combinations appear and yield

$$\sinh^4 x = \tfrac{1}{8} \cosh 4x - \tfrac{1}{2} \cosh 2x + \tfrac{3}{8}.$$

Now consider the relationship between this expression and the LHS of the given equation. They are clearly closely related; one is a multiple of the other, except in respect of the constant term. Making compensating corrections to the constant term allows us to rewrite the equation in terms of $\sinh^4 x$ as follows:

$$2 \cosh 4x - 8 \cosh 2x + (6 - 1) = 0,$$

$$16 \sinh^4 x - 1 = 0,$$

$$\sinh^4 x = \tfrac{1}{16},$$

$$\sinh x = \pm \tfrac{1}{2} \quad \text{(real solutions only)}.$$

We now use the explicit expression for the inverse hyperbolic sine, namely

$$\text{if } y = \sinh^{-1} z, \text{ then } y = \ln(\sqrt{1 + z^2} + z),$$

to give in this case

$$x = \ln\left(\sqrt{1 + \tfrac{1}{4}} \pm \tfrac{1}{2}\right) = 0.481 \text{ or } -0.481.$$

5.27 A closed barrel has as its curved surface the surface obtained by rotating about the x-axis the part of the curve

$$y = a[\,2 - \cosh(x/a)\,]$$

lying in the range $-b \le x \le b$, where $b < a \cosh^{-1} 2$. Show that the total surface area, A, of the barrel is given by

$$A = \pi a[\,9a - 8a\exp(-b/a) + a\exp(-2b/a) - 2b\,].$$

If s is the length of the curve defining the surface (measured from $x = 0$) then $ds^2 = dx^2 + dy^2$ and consequently $ds/dx = (1 + y'^2)^{1/2}$.

For this particular surface,

$$y = a\left(2 - \cosh\frac{x}{a}\right)$$

$$\text{and } \frac{dy}{dx} = -\sinh\frac{x}{a}.$$

It follows that

$$\frac{ds}{dx} = \left[1 + \left(\frac{dy}{dx}\right)^2\right]^{1/2}$$

$$= \left(1 + \sinh^2\frac{x}{a}\right)^{1/2}$$

$$= \cosh\frac{x}{a}.$$

The curved surface area, A_1, is given by

$$A_1 = 2\int_0^b 2\pi y\, ds$$

$$= 2\int_0^b 2\pi y\, \frac{ds}{dx}\, dx$$

$$= 4\pi a\int_0^b \left(2\cosh\frac{x}{a} - \cosh^2\frac{x}{a}\right) dx, \qquad \text{use } \cosh^2 z = \tfrac{1}{2}(\cosh 2z + 1)$$

$$= 4\pi a\int_0^b \left(2\cosh\frac{x}{a} - \frac{1}{2} - \frac{1}{2}\cosh\frac{2x}{a}\right) dx$$

$$= 4\pi a\left[2a\sinh\frac{x}{a} - \frac{x}{2} - \frac{a}{4}\sinh\frac{2x}{a}\right]_0^b$$

$$= \pi a\left(8a\sinh\frac{b}{a} - 2b - a\sinh\frac{2b}{a}\right).$$

The area, A_2, of the two flat ends is given by

$$A_2 = 2\pi a^2 \left(2 - \cosh\frac{b}{a}\right)^2$$

$$= 2\pi a^2 \left(4 - 4\cosh\frac{b}{a} + \cosh^2\frac{b}{a}\right).$$

And so the total area is

$$A = \pi a \left[4a\left(e^{b/a} - e^{-b/a}\right) - 2b - \frac{a}{2}\left(e^{2b/a} - e^{-2b/a}\right)\right.$$

$$\left. +8a - 4a\left(e^{b/a} + e^{-b/a}\right) + \frac{2a}{4}\left(e^{2b/a} + 2 + e^{-2b/a}\right)\right]$$

$$= \pi a \left(9a - 8ae^{-b/a} + ae^{-2b/a} - 2b\right).$$

6 Series and limits

6.1 Sum the even numbers between 1000 and 2000 inclusive.

We must first express the given sum in terms of a summation for which we have an explicit form. The result that is needed is clearly

$$S_N = \sum_{n=1}^{N} n = \frac{1}{2} N(N+1),$$

and we must rewrite the given summation in terms of sums of this form:

$$\sum_{n(\text{even})=1000}^{n=2000} n = \sum_{m=500}^{m=1000} 2m$$

$$= 2(S_{1000} - S_{499})$$

$$= 2 \left(\frac{1}{2} \times 1000 \times 1001 - \frac{1}{2} \times 499 \times 500 \right)$$

$$= 751\,500.$$

6.3 Prove that

$$\sum_{1}^{N} n(n+1)(n+2) = \tfrac{1}{4} N(N+1)(N+2)(N+3).$$

This problem can be approached in two ways: (i) using results already derived for the sums of the first N terms of the series $\sum n$, $\sum n^2$ and $\sum n^3$ or (ii) using proof by induction.

(i) We use the results $\sum n = \frac{1}{2} N(N+1)$, $\sum n^2 = \frac{1}{6} N(N+1)(2N+1)$ and $\sum n^3 = \frac{1}{4} N^2(N+1)^2$.

$$\sum_{1}^{N} n(n+1)(n+2) = \sum_{1}^{N} (n^3 + 3n^2 + 2n)$$

$$= \frac{1}{4} N^2(N+1)^2 + \frac{3}{6} N(N+1)(2N+1) + \frac{2}{2} N(N+1)$$

$$= \frac{1}{4} N(N+1)[N(N+1) + 2(2N+1) + 4]$$

$$= \frac{1}{4}N(N+1)[N^2 + 5N + 6]$$

$$= \frac{1}{4}N(N+1)(N+2)(N+3).$$

(ii) Assume that the result is true for $N = M$. Then, for $N = M + 1$

$$\sum_{1}^{M+1} n(n+1)(n+2) = \frac{1}{4}M(M+1)(M+2)(M+3) + (M+1)(M+2)(M+3)$$

$$= \frac{1}{4}(M+1)(M+2)(M+3)[M+4],$$

which is the assumed result, but with $N = M + 1$. Since $(1)(2)(3) = \frac{1}{4}(1)(2)(3)(4)$, the result is trivially true for $N = 1$, and it follows from the general method of proof by induction that the result is true for all finite N.

6.5 How does the convergence of the series

$$\sum_{n=r}^{\infty} \frac{(n-r)!}{n!}$$

depend on the integer r?

For $r \leq 1$, each term of the series is greater than or equal to the corresponding term of $\sum 1/n$, which is known to be divergent (for a proof, see any standard textbook). Thus, by the comparison test, the given series is also divergent.

For $r \geq 2$, each term of the series is less than or equal to the corresponding term of

$$\sum_{1}^{\infty} \frac{1}{n(n+1)}.$$

By writing this latter sum as

$$\sum_{n=1}^{\infty} \frac{1}{n(n+1)} = \sum_{n=1}^{\infty} \left(\frac{1}{n} - \frac{1}{n+1} \right)$$

$$= \left(1 - \frac{1}{2} \right) + \left(\frac{1}{2} - \frac{1}{3} \right) + \left(\frac{1}{3} - \frac{1}{4} \right) + \cdots$$

$$= 1 + \left(-\frac{1}{2} + \frac{1}{2} \right) + \left(-\frac{1}{3} + \frac{1}{3} \right) + \cdots \to 1,$$

it is shown to be convergent. Thus, by the comparison test, the given series is also convergent when $r \geq 2$.

6.7 Find the sum, S_N, of the first N terms of the following series and hence determine whether the series are convergent, divergent or oscillatory:

(a) $\sum_{n=1}^{\infty} \ln \left(\frac{n+1}{n} \right)$,   (b) $\sum_{n=0}^{\infty} (-2)^n$,   (c) $\sum_{n=1}^{\infty} \frac{(-1)^{n+1} n}{3^n}$.

(a) We express this series as the difference between two series with similar terms and find that the terms cancel in pairs, leaving an explicit expression that contains only the last term of the first series and the first term of the second:

$$\sum_{n=1}^{N} \ln \frac{n+1}{n} = \sum_{n=1}^{N} \ln(n+1) - \sum_{n=1}^{N} \ln n = \ln(N+1) - \ln 1.$$

As $\ln(N+1) \to \infty$ as $N \to \infty$, the series diverges.

   (b) Applying the normal formula for a geometric sum gives

$$\sum_{n=0}^{N-1} (-2)^n = \frac{1 - (-2)^N}{3}.$$

The series therefore oscillates infinitely.

   (c) Denote the partial sum by S_N. Then,

$$S_N = \sum_{n=1}^{N} \frac{(-1)^{n+1} n}{3^n},$$

$$\frac{1}{3} S_N = \sum_{n=1}^{N} \frac{(-1)^{n+1} n}{3^{n+1}} = \sum_{s=2}^{N+1} \frac{(-1)^s (s-1)}{3^s}$$

$$= \sum_{s=2}^{N+1} \frac{(-1)^s s}{3^s} - \sum_{s=2}^{N+1} \frac{(-1)^s}{3^s}.$$

Separating off the last term of the first series on the RHS and adding S_N to both sides, with the S_N added to the RHS having its $n = 1$ term written explicitly, yields

$$\frac{4}{3} S_N = \frac{(-1)^2 1}{3} + \sum_{n=2}^{N} \frac{(-1)^{n+1} n}{3^n} + \sum_{s=2}^{N} \frac{(-1)^s s}{3^s}$$

$$+ \frac{(-1)^{N+1}(N+1)}{3^{N+1}} - \sum_{s=2}^{N+1} \frac{(-1)^s}{3^s}$$

$$= \frac{1}{3} + \frac{(-1)^{N+1}(N+1)}{3^{N+1}} - \frac{1}{9} \frac{1 - (-\frac{1}{3})^N}{1 - (-\frac{1}{3})}.$$

To obtain the last line we note that on the RHS the second and third terms (both summations) cancel and that the final term is a geometric series (with leading term $-\frac{1}{9}$). This result can be rearranged as

$$S_N = \frac{3}{16} \left[1 - \left(-\frac{1}{3}\right)^N \right] + \frac{3N}{4} \left(-\frac{1}{3}\right)^{N+1},$$

from which it is clear that the series converges to a sum of $\frac{3}{16}$.

6.9 Use the difference method to sum the series

$$\sum_{n=2}^{N} \frac{2n-1}{2n^2(n-1)^2}.$$

We try to write the nth term as the difference between two consecutive values of a partial-fraction function of n. Since the second power of n appears in the denominator the function will need two terms, An^{-2} and Bn^{-1}. Hence, we must have

$$\frac{2n-1}{2n^2(n-1)^2} = \frac{A}{n^2} + \frac{B}{n} - \left[\frac{A}{(n-1)^2} + \frac{B}{n-1} \right]$$

$$= \frac{A[-2n+1] + B[n(n-1)(n-1-n)]}{n^2(n-1)^2}.$$

The powers of n in the numerators can be equated consistently if we take $A = -\frac{1}{2}$ and $B = 0$. Thus:

$$\frac{2n-1}{2n^2(n-1)^2} = \frac{1}{2} \left[\frac{1}{(n-1)^2} - \frac{1}{n^2} \right].$$

We can now carry out the summation, in which the second component of each pair of terms cancels the first component of the next pair, leaving only the initial and very final components:

$$\sum_{n=2}^{N} \frac{2n-1}{2n^2(n-1)^2} = \frac{1}{2} \sum_{n=2}^{N} \left[\frac{1}{(n-1)^2} - \frac{1}{n^2} \right]$$

$$= \frac{1}{2} \left(\frac{1}{1} - \frac{1}{N^2} \right)$$

$$= \tfrac{1}{2}(1 - N^{-2}).$$

6.11 Prove that

$$\cos\theta + \cos(\theta + \alpha) + \cdots + \cos(\theta + n\alpha) = \frac{\sin\frac{1}{2}(n+1)\alpha}{\sin\frac{1}{2}\alpha} \cos(\theta + \tfrac{1}{2}n\alpha).$$

From de Moivre's theorem, the required sum, S, is the real part of the sum of the geometric series $\sum_{r=0}^{n} e^{i\theta} e^{ir\alpha}$. Using the formula for the partial sum of a geometric series, and multiplying by a factor that makes the denominator real, we have

$$S = \text{Re}\left(e^{i\theta} \frac{1 - e^{i(n+1)\alpha}}{1 - e^{i\alpha}} \frac{1 - e^{-i\alpha}}{1 - e^{-i\alpha}} \right)$$

$$= \frac{\cos\theta - \cos[(n+1)\alpha + \theta] - \cos(\theta - \alpha) + \cos(\theta + n\alpha)}{2 \times 2\sin^2\frac{1}{2}\alpha}$$

$$= \frac{2\sin(\theta - \frac{1}{2}\alpha)\sin(-\frac{1}{2}\alpha) + 2\sin(n\alpha + \frac{1}{2}\alpha + \theta)\sin\frac{1}{2}\alpha}{4\sin^2\frac{1}{2}\alpha}$$

$$= \frac{2\sin\frac{1}{2}\alpha\, 2\cos(\frac{1}{2}n\alpha + \theta)\,\sin[\frac{1}{2}(n+1)\alpha]}{4\sin^2\frac{1}{2}\alpha}$$

$$= \frac{\sin\frac{1}{2}(n+1)\alpha}{\sin\frac{1}{2}\alpha}\cos(\theta + \frac{1}{2}n\alpha).$$

In the course of this manipulation we have used the identity $1 - \cos\theta = 2\sin^2\frac{1}{2}\theta$ and the formulae for $\cos A - \cos B$ and $\sin A - \sin B$.

6.13 Find the real values of x for which the following series are convergent:

(a) $\displaystyle\sum_{n=1}^{\infty}\frac{x^n}{n+1}$, (b) $\displaystyle\sum_{n=1}^{\infty}(\sin x)^n$, (c) $\displaystyle\sum_{n=1}^{\infty}n^x$, (d) $\displaystyle\sum_{n=1}^{\infty}e^{nx}$.

(a) Using the ratio test:

$$\lim_{n\to\infty}\frac{u_{n+1}}{u_n} = \lim_{n\to\infty}\frac{x^{n+1}}{n+2}\frac{n+1}{x^n} = x.$$

Thus the series is convergent for all $|x| < 1$. At $x = 1$ the series diverges, as shown in any standard text, whilst at $x = -1$ it converges by the alternating series test. Thus we have convergence for $-1 \le x < 1$.

 (b) For all x other than $x = (2m \pm \frac{1}{2})\pi$, where m is an integer, $|\sin x| < 1$ and so convergence is assured by the ratio test. At $x = (2m + \frac{1}{2})\pi$ the series diverges, whilst at $x = (2m - \frac{1}{2})\pi$ it oscillates finitely.

 (c) This is the Riemann zeta series with p written as $-x$. Thus the series converges for all $x < -1$.

 (d) The ratio of successive terms is e^x (independent of n) and for this to be less than unity in magnitude requires x to be negative. Thus the series is convergent when $x < 0$.

6.15 Determine whether the following series are absolutely convergent, convergent or oscillatory:

(a) $\displaystyle\sum_{n=1}^{\infty}\frac{(-1)^n}{n^{5/2}}$, (b) $\displaystyle\sum_{n=1}^{\infty}\frac{(-1)^n(2n+1)}{n}$, (c) $\displaystyle\sum_{n=0}^{\infty}\frac{(-1)^n|x|^n}{n!}$,

(d) $\displaystyle\sum_{n=0}^{\infty}\frac{(-1)^n}{n^2 + 3n + 2}$, (e) $\displaystyle\sum_{n=1}^{\infty}\frac{(-1)^n 2^n}{n^{1/2}}$.

(a) The sum $\sum n^{-5/2}$ is convergent (by comparison with $\sum n^{-2}$) and so the series $\sum(-1)^n n^{-5/2}$ is absolutely convergent.

(b) The magnitude of the individual terms $\to 2$ and not to zero; thus the series cannot converge. In fact it oscillates finitely about the value $-(1 + \ln 2)$.

(c) The magnitude of the successive-term ratio is

$$\left| \frac{u_{n+1}}{u_n} \right| = \frac{|x|^{n+1}}{(n+1)!} \frac{n!}{|x|^n} = \frac{|x|}{n} \to 0 \quad \text{for all } x.$$

Thus, the series is absolutely convergent for all finite x.

(d) The polynomial in the denominator has all positive signs and a non-zero constant term; it is therefore always strictly positive. Thus, to test for absolute convergence, we need to replace the numerator by its absolute value and consider $\sum_{n=0}^{N}(n^2 + 3n + 2)^{-1}$:

$$\sum_{n=0}^{N} \frac{1}{n^2 + 3n + 2} = \sum_{n=0}^{N} \left(\frac{1}{n+1} - \frac{1}{n+2} \right) = 1 - \frac{1}{N+2} \to 1 \quad \text{as} \quad N \to \infty.$$

Thus the given series is absolutely convergent.

(e) The magnitude of the individual terms does not tend to zero; in fact, it grows monotonically. The effect of the alternating signs is to make the series oscillate infinitely.

6.17 Prove that

$$\sum_{n=2}^{\infty} \ln \left[\frac{n^r + (-1)^n}{n^r} \right]$$

is absolutely convergent for $r = 2$, but only conditionally convergent for $r = 1$.

In each case divide the sum into two sums, one for n even and one for n odd.

(i) For $r = 2$, consider first the even series:

$$\sum_{n \text{ even}} \ln \left(\frac{n^2 + 1}{n^2} \right) = \sum_{n \text{ even}} \ln \left(1 + \frac{1}{n^2} \right)$$

$$= \sum_{n \text{ even}} \left(\frac{1}{n^2} - \frac{1}{2n^4} + \cdots \right).$$

The nth logarithmic term is positive for all n but, as shown above, less than n^{-2}. It follows from the comparison test that the series is (absolutely) convergent.

For the odd series we consider

$$\ln \left[\frac{(2m+1)^2 - 1}{(2m+1)^2} \right] = \ln \frac{4m^2 + 4m}{4m^2 + 4m + 1}$$

$$= -\ln \left[1 + \frac{1}{4m(m+1)} \right].$$

By a similar argument to that above, each term is negative but greater than $-[4m(m+1)]^{-1}$. Again, the comparison test shows that the series is (absolutely) convergent.

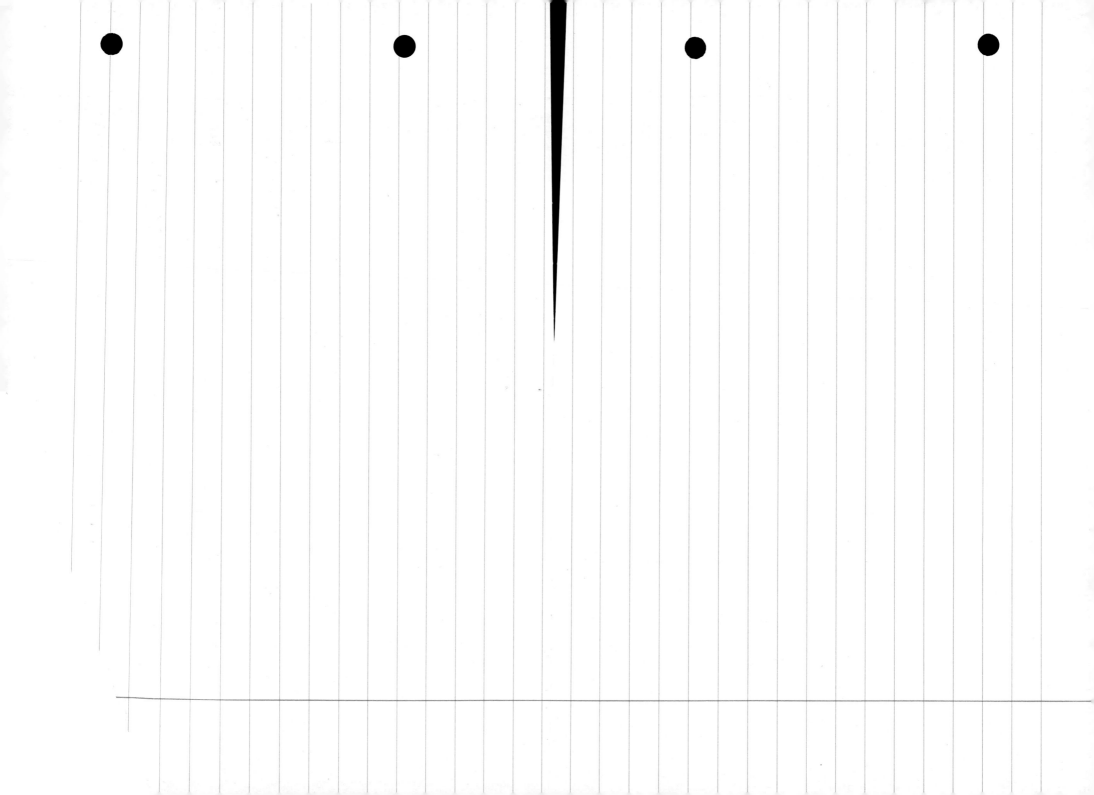

Weight 19.6 N   hill

Initial energy = ⊕ mgh →

⊕ W = mg   m = $\frac{W}{g}$ = $\frac{19.86}{9.8}$ = 2

Initial energy = 2 × 9.8 × 16 = 313.6 J

Final energy → kinetic energy

313.6 = $\frac{1}{2}mv^2$

627.2 = mv²   m = 2   so

313.6 = √²

v = 17.7087549

≈ 17.7 ms⁻¹

Thus the original series, being the sum of two absolutely convergent series, is also absolutely convergent.

(ii) For $r = 1$ we have to consider $\ln[(n \pm 1)/n]$, whose expansion contains a term $\pm n^{-1}$ and other inverse powers of n. The summations over the other powers converge and cannot cancel the divergence arising from $\sum \pm n^{-1}$. Thus both the even and odd series diverge; consequently, the original series cannot be absolutely convergent.

However, if we group together consecutive pairs of terms, $n = 2m$ and $n = 2m + 1$, then we see that

$$\sum_{n=2}^{\infty} \ln \left[\frac{n + (-1)^n}{n} \right] = \sum_{m=1}^{\infty} \left[\ln \frac{2m + 1}{2m} + \ln \frac{2m + 1 - 1}{2m + 1} \right]$$

$$= \sum_{m=1}^{\infty} \ln 1 = \sum_{m=1}^{\infty} 0 = 0,$$

i.e. the terms cancel in pairs and the series is conditionally convergent to zero.

6.19 Demonstrate that rearranging the order of its terms can make a conditionally convergent series converge to a different limit by considering the series $\sum (-1)^{n+1} n^{-1} = \ln 2 = 0.693$. Rearrange the series as

$$S = \tfrac{1}{1} + \tfrac{1}{3} - \tfrac{1}{2} + \tfrac{1}{5} + \tfrac{1}{7} - \tfrac{1}{4} + \tfrac{1}{9} + \tfrac{1}{11} - \tfrac{1}{6} + \tfrac{1}{13} + \cdots$$

and group each set of three successive terms. Show that the series can then be written

$$\sum_{m=1}^{\infty} \frac{8m - 3}{2m(4m - 3)(4m - 1)},$$

which is convergent (by comparison with $\sum n^{-2}$) and contains only positive terms. Evaluate the first of these and hence deduce that S is not equal to $\ln 2$.

Proceeding as indicated, we have

$$S = \left(\frac{1}{1} + \frac{1}{3} - \frac{1}{2} \right) + \left(\frac{1}{5} + \frac{1}{7} - \frac{1}{4} \right) + \left(\frac{1}{9} + \frac{1}{11} - \frac{1}{6} \right) + \cdots$$

$$= \sum_{m=1}^{\infty} \left(\frac{1}{4m - 3} + \frac{1}{4m - 1} - \frac{1}{2m} \right)$$

$$= \sum_{m=1}^{\infty} \frac{(8m^2 - 2m) + (8m^2 - 6m) - (16m^2 - 16m + 3)}{2m(4m - 3)(4m - 1)}$$

$$= \sum_{m=1}^{\infty} \frac{8m - 3}{2m(4m - 3)(4m - 1)}.$$

As noted, this series is convergent and contains only positive terms. The first of these terms ($m = 1$) is $5/6 = 0.833$. This, by itself, is greater than the known sum (0.693) of the original series. Thus S cannot be equal to $\ln 2$.

6.21 A Fabry–Pérot interferometer consists of two parallel heavily silvered glass plates; light enters normally to the plates and undergoes repeated reflections between them, with a small transmitted fraction emerging at each reflection. Find the intensity $|B|^2$ of the emerging wave, where

$$B = A(1 - r) \sum_{n=0}^{\infty} r^n e^{in\phi},$$

with r and ϕ real.

This is a simple geometric series but with a complex common ratio $re^{i\phi}$. Thus we have

$$B = A(1 - r) \sum_{n=0}^{\infty} r^n e^{in\phi}$$

$$= A \frac{1 - r}{1 - re^{i\phi}}.$$

To obtain the intensity $|B|^2$ we multiply this result by its complex conjugate, recalling that r and ϕ are real, but A may not be:

$$|B|^2 = \frac{|A|^2(1 - r)^2}{(1 - re^{i\phi})(1 - re^{-i\phi})}$$

$$= \frac{|A|^2(1 - r)^2}{1 - 2r \cos\phi + r^2}.$$

6.23 Starting from the Maclaurin series for $\cos x$, show that

$$(\cos x)^{-2} = 1 + x^2 + \frac{2x^4}{3} + \cdots$$

Deduce the first three terms in the Maclaurin series for $\tan x$.

From the Maclaurin series for (or definition of) $\cos x$,

$$\cos x = 1 - \frac{x^2}{2!} + \frac{x^4}{4!} + \cdots$$

Using the binomial expansion of $(1 + z)^{-2}$, we have

$$(\cos x)^{-2} = \left(1 - \frac{x^2}{2!} + \frac{x^4}{4!} + \cdots\right)^{-2}$$

$$= 1 - 2\left(-\frac{x^2}{2!} + \frac{x^4}{4!} + \cdots\right) + \frac{2 \cdot 3}{2!}\left(-\frac{x^2}{2!} + \frac{x^4}{4!} + \cdots\right)^2 + \cdots$$

$$= 1 + x^2 + x^4\left(-\frac{2}{4!} + \frac{2 \cdot 3}{2! \, 2! \, 2!}\right) + O(x^6)$$

$$= 1 + x^2 + \tfrac{2}{3}x^4 + \cdots$$

We now integrate both sides of the expansion from 0 to x, noting that $(\cos x)^{-2} \equiv \sec^2 x$ and that this integrates to $\tan x$. Thus:

$$\tan x = \int_0^x \sec^2 u \, du = x + \frac{x^3}{3} + \frac{2x^5}{15} + \cdots$$

6.25 Writing the nth derivative of $f(x) = \sinh^{-1} x$ as

$$f^{(n)}(x) = \frac{P_n(x)}{(1+x^2)^{n-1/2}},$$

where $P_n(x)$ is a polynomial (of degree $n-1$), show that the $P_n(x)$ satisfy the recurrence relation

$$P_{n+1}(x) = (1+x^2)P_n'(x) - (2n-1)x P_n(x).$$

Hence generate the coefficients necessary to express $\sinh^{-1} x$ as a Maclaurin series up to terms in x^5.

With $f(x) = \sinh^{-1} x$,

$$x = \sinh f \quad \Rightarrow \quad \frac{dx}{df} = \cosh f \quad \Rightarrow \quad \frac{df}{dx} = \frac{1}{\cosh f} = \frac{1}{(1+x^2)^{1/2}}.$$

Thus $P_1(x) = 1$; we will need this as a starting value for the recurrence relation.
    With the definition of $P_n(x)$ given,

$$f^{(n)} = \frac{P_n}{(1+x^2)^{n-1/2}},$$

$$f^{(n+1)} = \frac{P_n'}{(1+x^2)^{n-1/2}} - \frac{(n-\frac{1}{2})2x \, P_n}{(1+x^2)^{n+1/2}}$$

$$= \frac{(1+x^2)P_n' - (2n-1)x P_n}{(1+x^2)^{n+1-1/2}}.$$

It then follows that

$$P_{n+1}(x) = (1+x^2)P_n'(x) - (2n-1)x P_n(x).$$

With $P_1 = 1$, as shown,

$$P_2 = (1+x^2)0 - (2-1)x \, 1 = -x,$$
$$P_3 = (1+x^2)(-1) - (4-1)x(-x) = 2x^2 - 1,$$
$$P_4 = (1+x^2)(4x) - (6-1)x(2x^2 - 1) = 9x - 6x^3,$$
$$P_5 = (1+x^2)(9 - 18x^2) - (8-1)x(9x - 6x^3) = 24x^4 - 72x^2 + 9.$$

The corresponding values of $f^{(n)}(0) = P_n(0)/(1 + 0^2)^{n-1/2}$ can then be used to express the Maclaurin series for $\sinh^{-1} x$ as

$$\sinh^{-1} x = f(0) + \sum_{n=1}^{\infty} \frac{f^n(0)x^n}{n!} = x - \frac{x^3}{3!} + \frac{9x^5}{5!} - \cdots$$

6.27 By using the logarithmic series, prove that if a and b are positive and nearly equal then

$$\ln \frac{a}{b} \simeq \frac{2(a - b)}{a + b}.$$

Show that the error in this approximation is about $2(a - b)^3/[3(a + b)^3]$.

Write $a + b = 2c$ and $a - b = 2\delta$. Then

$$\ln \frac{a}{b} = \ln a - \ln b$$

$$= \ln(c + \delta) - \ln(c - \delta)$$

$$= \ln c + \ln\left(1 + \frac{\delta}{c}\right) - \ln c - \ln\left(1 - \frac{\delta}{c}\right)$$

$$= \left(\frac{\delta}{c} - \frac{\delta^2}{2c^2} + \frac{\delta^3}{3c^3} - \cdots\right) - \left(-\frac{\delta}{c} - \frac{\delta^2}{2c^2} - \frac{\delta^3}{3c^3} - \cdots\right)$$

$$= \frac{2\delta}{c} + \frac{2}{3}\left(\frac{\delta}{c}\right)^3 + \cdots$$

$$= \frac{2(a - b)}{a + b} + \frac{2}{3}\left(\frac{a - b}{a + b}\right)^3 + \cdots,$$

i.e. as stated in the question.

We note that other approximations are possible, and equally valid, e.g. setting $b = a + \epsilon$ leading to $-(\epsilon/a)(1 - \epsilon/2a + \epsilon^2/3a^2 - \cdots)$, but the given one, expanding symmetrically about $c = (a + b)/2$, contains no quadratic terms in $(a - b)$, only cubic and higher terms.

6.29 Find the limit as $x \to 0$ of $[\sqrt{1 + x^m} - \sqrt{1 - x^m}]/x^n$, in which m and n are positive integers.

Using the binomial expansions of the terms in the numerator,

$$\frac{\sqrt{1 + x^m} - \sqrt{1 - x^m}}{x^n} = \frac{1 + \frac{1}{2}x^m + \cdots - (1 - \frac{1}{2}x^m + \cdots)}{x^n}$$

$$= \frac{x^m + \cdots}{x^n}$$

$$= x^{m-n} + \cdots$$

Thus the limit of the function as $x \to 0$ is 0 for $m > n$, 1 for $m = n$ and ∞ for $m < n$.

6.31 Find the limits of the following functions:

(a) $\dfrac{x^3 + x^2 - 5x - 2}{2x^3 - 7x^2 + 4x + 4}$, as $x \to 0, x \to \infty$ and $x \to 2$;

(b) $\dfrac{\sin x - x \cosh x}{\sinh x - x}$, as $x \to 0$;

(c) $\displaystyle\int_x^{\pi/2} \left(\dfrac{y \cos y - \sin y}{y^2} \right) dy$, as $x \to 0$.

(a) Denote the ratio of polynomials by $f(x)$. Then

$$\lim_{x \to 0} f(x) = \lim_{x \to 0} \frac{x^3 + x^2 - 5x - 2}{2x^3 - 7x^2 + 4x + 4} = \frac{-2}{4} = -\frac{1}{2};$$

$$\lim_{x \to \infty} f(x) = \lim_{x \to \infty} \frac{1 + x^{-1} - 5x^{-2} - 2x^{-3}}{2 - 7x^{-1} + 4x^{-2} + 4x^{-3}} = \frac{1}{2};$$

$$\lim_{x \to 2} f(x) = \lim_{x \to 2} \frac{x^3 + x^2 - 5x - 2}{2x^3 - 7x^2 + 4x + 4} = \frac{0}{0}.$$

This final value is indeterminate and so, using l'Hôpital's rule, consider instead

$$\lim_{x \to 2} f(x) = \lim_{x \to 2} \frac{3x^2 + 2x - 5}{6x^2 - 14x + 4} = \frac{11}{0} = \infty.$$

(b) Using l'Hôpital's rule repeatedly,

$$\lim_{x \to 0} \frac{\sin x - x \cosh x}{\sinh x - x} = \lim_{x \to 0} \frac{\cos x - \cosh x - x \sinh x}{\cosh x - 1}$$

$$= \lim_{x \to 0} \frac{-\sin x - \sinh x - \sinh x - x \cosh x}{\sinh x}$$

$$= \lim_{x \to 0} \frac{-\cos x - 2 \cosh x - \cosh x - x \sinh x}{\cosh x} = -4.$$

(c) Before taking the limit we need to find a closed form for the integral. So,

$$\lim_{x \to 0} \int_x^{\pi/2} \left(\frac{y \cos y - \sin y}{y^2} \right) dy = \lim_{x \to 0} \int_x^{\pi/2} \frac{d}{dy} \left(\frac{\sin y}{y} \right) dy$$

$$= \lim_{x \to 0} \left[\frac{\sin y}{y} \right]_x^{\pi/2}$$

$$= \lim_{x \to 0} \left(\frac{2}{\pi} - \frac{\sin x}{x} \right)$$

$$= \lim_{x \to 0} \left[\frac{2}{\pi} - \frac{1}{x} \left(x - \frac{x^3}{3!} + \cdots \right) \right]$$

$$= \frac{2}{\pi} - 1.$$

6.33 Using a first-order Taylor expansion about $x = x_0$, show that a better approximation than x_0 to the solution of the equation

$$f(x) = \sin x + \tan x = 2$$

is given by $x = x_0 + \delta$, where

$$\delta = \frac{2 - f(x_0)}{\cos x_0 + \sec^2 x_0}.$$

(a) Use this procedure twice to find the solution of $f(x) = 2$ to six significant figures, given that it is close to $x = 0.9$.

(b) Use the result in (a) to deduce, to the same degree of accuracy, one solution of the quartic equation

$$y^4 - 4y^3 + 4y^2 + 4y - 4 = 0.$$

(a) We write the solution to $f(x) = \sin x + \tan x = 2$ as $x = x_0 + \delta$. Substituting this form and retaining the first-order terms in δ in the Taylor expansions of $\sin x$ and $\tan x$ we obtain

$$\sin x_0 + \delta \cos x_0 + \cdots + \tan x_0 + \delta \sec^2 x_0 + \cdots = 2$$

$$\delta = \frac{2 - \sin x_0 - \tan x_0}{\cos x_0 + \sec^2 x_0}.$$

With $x_0 = 0.9$,

$$\delta_1 = \frac{2 - 0.783327 - 1.260158}{0.621610 + 2.587999} = \frac{-0.043485}{3.209609} = -0.013548,$$

making the first improved approximation $x_1 = x_0 + \delta_1 = 0.886452$.

Now, using x_1 instead of x_0 and repeating the process gives

$$\delta_2 = \frac{2 - 0.774833 - 1.225682}{0.632165 + 2.502295} = \frac{-5.15007 \times 10^{-4}}{3.13446} = -1.6430 \times 10^{-4},$$

making the second improved approximation $x_2 = x_1 + \delta_2 = 0.886287$. The method used up to here does not *prove* that this latest answer is accurate to six significant figures, but a further application of the procedure shows that $\delta_3 \approx 3 \times 10^{-7}$.

(b) In order to make use of the result in part (a) we need to make a change of variable that converts the geometric equation into an algebraic one. Since $\tan x$ can be expressed in terms of $\sin x$, if we set $y = \sin x$ in the equation $\sin x + \tan x = 2$, it will become an algebraic equation:

$$\sin x + \tan x = \sin x + \frac{\sin x}{\cos x} = 2,$$

$$\Rightarrow \quad y + \frac{y}{\sqrt{1 - y^2}} = 2,$$

$$\frac{y^2}{1 - y^2} = (2 - y)^2,$$

$$y^2 = (1 - y^2)(4 - 4y + y^2)$$
$$= -y^4 + 4y^3 - 3y^2 - 4y + 4,$$
$$0 = y^4 - 4y^3 + 4y^2 + 4y - 4.$$

This is the equation that is to be solved. Thus, since $x = 0.886287$ is an approximation to the solution of $\sin x + \tan x = 2$, $y = \sin x = 0.774730$ is an approximation to one of the solutions of $y^4 - 4y^3 + 4y^2 + 4y - 4 = 0$ to the same degree of accuracy.

We note that an equally plausible change of variable is to set $y = \tan x$, with $\sin x$ expressed as $\tan x / \sec x$, i.e. as $y/\sqrt{1 + y^2}$. With this substitution the resulting algebraic equation is the quartic $y^4 - 4y^3 + 4y^2 - 4y + 4 = 0$ (very similar to, but not exactly the same as, the given quartic equation). The reader may wish to verify this. By a parallel argument to that above, $y = \tan 0.886287 = 1.225270$ is an approximate solution of this second quartic equation.

6.35 In quantum theory, a system of oscillators, each of fundamental frequency v and interacting at temperature T, has an average energy $\bar{E}$ given by

$$\bar{E} = \frac{\sum_{n=0}^{\infty} nhve^{-nx}}{\sum_{n=0}^{\infty} e^{-nx}},$$

where $x = hv/kT$, h and k being the Planck and Boltzmann constants, respectively. Prove that both series converge, evaluate their sums, and show that at high temperatures $\bar{E} \approx kT$, whilst at low temperatures $\bar{E} \approx hv \exp(-hv/kT)$.

In the expression

$$\bar{E} = \frac{\sum_{n=0}^{\infty} nhve^{-nx}}{\sum_{n=0}^{\infty} e^{-nx}},$$

the ratio of successive terms in the series in the numerator is given by

$$\left| \frac{a_{n+1}}{a_n} \right| = \left| \frac{(n + 1)hve^{-(n+1)x}}{nhve^{-nx}} \right| = \left| \frac{n + 1}{n} e^{-x} \right| \to e^{-x} \quad \text{as} \quad n \to \infty,$$

where $x = hv/kT$. Since $x > 0$, $e^{-x} < 1$, and the series is convergent by the ratio test.

The series in the denominator is a geometric series with common ratio $r = e^{-x}$. This is <1 and so the series converges with sum

$$S(x) = 1 + e^{-x} + e^{-2x} + \cdots + e^{-nx} + \cdots = \frac{1}{1 - e^{-x}}.$$

Now consider

$$-\frac{dS(x)}{dx} = e^{-x} + 2e^{-2x} + \cdots + ne^{-nx} + \cdots$$

The series on the RHS, when multiplied by $h\nu$, gives the numerator in the expression for $\bar{E}$; the numerator therefore has the value

$$-\frac{dS(x)}{dx} = -\frac{d}{dx}\left(\frac{1}{1-e^{-x}}\right) = \frac{e^{-x}}{(1-e^{-x})^2}.$$

Hence,

$$\bar{E} = \frac{h\nu\,e^{-x}}{(1-e^{-x})^2}\frac{1-e^{-x}}{1} = \frac{h\nu}{e^x-1}.$$

At high temperatures, $x \ll 1$ and

$$\bar{E} = \frac{h\nu}{\left(1+\frac{h\nu}{kT}+\cdots\right)-1} \approx kT.$$

At low temperatures, $x \gg 1$ and $e^x \gg 1$. Thus the -1 in the denominator can be neglected and $\bar{E} \approx h\nu \exp(-h\nu/kT)$.

6.37 One of the factors contributing to the high relative permittivity of water to static electric fields is the permanent electric dipole moment, p, of the water molecule. In an external field E the dipoles tend to line up with the field, but they do not do so completely because of thermal agitation corresponding to the temperature, T, of the water. A classical (non-quantum) calculation using the Boltzmann distribution shows that the average polarisability per molecule, α, is given by

$$\alpha = \frac{p}{E}(\coth x - x^{-1}),$$

where $x = pE/(kT)$ and k is the Boltzmann constant.

At ordinary temperatures, even with high field strengths (10^4 V m^{-1} or more), $x \ll 1$. By making suitable series expansions of the hyperbolic functions involved, show that $\alpha = p^2/(3kT)$ to an accuracy of about one part in $15x^{-2}$.

As $x \ll 1$, we have to deal with a function that is the difference between two terms that individually tend to infinity as $x \to 0$. We will need to expand each in a series and consider the leading non-cancelling terms. The coth function will have to be expressed in terms of the series for the sinh and cosh functions, as follows:

$$\alpha = \frac{p}{E}\left(\coth x - \frac{1}{x}\right), \quad \text{with} \quad x = \frac{pE}{kT},$$

$$= \frac{p}{E}\left(\frac{\cosh x}{\sinh x} - \frac{1}{x}\right)$$

$$= \frac{p}{E}\left[\frac{1+\frac{x^2}{2!}+\frac{x^4}{4!}+\cdots}{x\left(1+\frac{x^2}{3!}+\frac{x^4}{5!}+\cdots\right)} - \frac{1}{x}\right]$$

$$= \frac{p}{Ex}\left\{\left(1+\frac{x^2}{2!}+\frac{x^4}{4!}+\cdots\right)\left[1-\left(\frac{x^2}{3!}+\frac{x^4}{5!}+\cdots\right)\right.\right.$$

$$\left.\left. +\left(\frac{x^2}{3!}+\frac{x^4}{5!}+\cdots\right)^2+\cdots\right]-1\right\}$$

$$= \frac{p}{Ex}\left[0 + x^2\left(\frac{1}{2!} - \frac{1}{3!}\right) + x^4\left(-\frac{1}{5!} + \frac{1}{(3!)^2} - \frac{1}{2!\,3!} + \frac{1}{4!}\right) + \cdots\right]$$

$$= \frac{px}{E}\left(\frac{1}{3} - \frac{x^2}{45} + \cdots\right).$$

Thus the polarisability $\approx px/3E = p^2/3kT$, with the correction term being a factor of about $x^2/15$ smaller.

7 Partial differentiation

7.1 Using the appropriate properties of ordinary derivatives, perform the following.

(a) Find all the first partial derivatives of the following functions $f(x, y)$:
  (i) $x^2 y$,   (ii) $x^2 + y^2 + 4$,   (iii) $\sin(x/y)$,   (iv) $\tan^{-1}(y/x)$,
  (v) $r(x, y, z) = (x^2 + y^2 + z^2)^{1/2}$.

(b) For (i), (ii) and (v), find $\partial^2 f/\partial x^2$, $\partial^2 f/\partial y^2$, and $\partial^2 f/\partial x \partial y$.

(c) For (iv) verify that $\partial^2 f/\partial x \partial y = \partial^2 f/\partial y \partial x$.

These are all straightforward applications of the definitions of partial derivatives.

(a) (i) $\dfrac{\partial f}{\partial x} = \dfrac{\partial(x^2 y)}{\partial x} = 2xy;$ $\dfrac{\partial f}{\partial y} = \dfrac{\partial(x^2 y)}{\partial y} = x^2.$

  (ii) $\dfrac{\partial f}{\partial x} = \dfrac{\partial(x^2 + y^2 + 4)}{\partial x} = 2x;$ $\dfrac{\partial f}{\partial y} = \dfrac{\partial(x^2 + y^2 + 4)}{\partial y} = 2y.$

  (iii) $\dfrac{\partial f}{\partial x} = \dfrac{\partial}{\partial x} \sin\left(\dfrac{x}{y}\right) = \cos\left(\dfrac{x}{y}\right) \dfrac{1}{y};$

  $\dfrac{\partial f}{\partial y} = \dfrac{\partial}{\partial y} \sin\left(\dfrac{x}{y}\right) = \cos\left(\dfrac{x}{y}\right) \dfrac{-x}{y^2}.$

  (iv) $\dfrac{\partial f}{\partial x} = \dfrac{\partial}{\partial x} \left[\tan^{-1}\left(\dfrac{y}{x}\right)\right] = \dfrac{1}{1 + \frac{y^2}{x^2}} \dfrac{-y}{x^2} = -\dfrac{y}{x^2 + y^2};$

  $\dfrac{\partial f}{\partial y} = \dfrac{\partial}{\partial y} \left[\tan^{-1}\left(\dfrac{y}{x}\right)\right] = \dfrac{1}{1 + \frac{y^2}{x^2}} \dfrac{1}{x} = \dfrac{x}{x^2 + y^2}.$

  (v) $\dfrac{\partial r}{\partial x} = \dfrac{\partial(x^2 + y^2 + z^2)^{1/2}}{\partial x} = \dfrac{\frac{1}{2} \times 2x}{(x^2 + y^2 + z^2)^{1/2}} = \dfrac{x}{r};$

  similarly for $\dfrac{\partial r}{\partial y}$ and $\dfrac{\partial r}{\partial z}$.

(b) (i) $\dfrac{\partial^2(x^2 y)}{\partial x^2} = \dfrac{\partial(2xy)}{\partial x} = 2y;$ $\dfrac{\partial^2(x^2 y)}{\partial y^2} = \dfrac{\partial(x^2)}{\partial y} = 0;$

  $\dfrac{\partial^2(x^2 y)}{\partial x \partial y} = \dfrac{\partial(x^2)}{\partial x} = 2x.$

(ii) $\dfrac{\partial^2(x^2 + y^2 + 4)}{\partial x^2} = \dfrac{\partial(2x)}{\partial x} = 2;$ $\quad$ $\dfrac{\partial^2(x^2 + y^2 + 4)}{\partial y^2} = \dfrac{\partial(2y)}{\partial y} = 2;$

$$\dfrac{\partial^2(x^2 + y^2 + 4)}{\partial x \partial y} = \dfrac{\partial(2y)}{\partial x} = 0.$$

(v) $\dfrac{\partial^2(x^2 + y^2 + z^2)^{1/2}}{\partial x^2} = \dfrac{\partial}{\partial x}\left(\dfrac{x}{r}\right) = \dfrac{1}{r} - \dfrac{x}{r^2}\dfrac{\partial r}{\partial x}$

$$= \dfrac{1}{r} - \dfrac{x}{r^2}\dfrac{x}{r} = \dfrac{y^2 + z^2}{r^3};$$

similarly for $\dfrac{\partial^2 r}{\partial y^2};$

$$\dfrac{\partial^2(x^2 + y^2 + z^2)^{1/2}}{\partial x \partial y} = \dfrac{\partial}{\partial x}\left(\dfrac{y}{r}\right) = -\dfrac{y}{r^2}\dfrac{x}{r} = -\dfrac{xy}{r^3}.$$

(c) $\dfrac{\partial^2 f}{\partial y \partial x} = \dfrac{\partial}{\partial y}\left(\dfrac{-y}{x^2 + y^2}\right) = -\dfrac{(x^2 + y^2) - y\,2y}{(x^2 + y^2)^2} = \dfrac{y^2 - x^2}{(x^2 + y^2)^2}$

and

$$\dfrac{\partial^2 f}{\partial x \partial y} = \dfrac{\partial}{\partial x}\left(\dfrac{x}{x^2 + y^2}\right) = \dfrac{(x^2 + y^2) - x\,2x}{(x^2 + y^2)^2} = \dfrac{y^2 - x^2}{(x^2 + y^2)^2},$$

thus verifying the general result for this particular case.

7.3 Show that the differential

$$df = x^2\,dy - (y^2 + xy)\,dx$$

is not exact, but that $dg = (xy^2)^{-1}df$ is exact.

If $df = A\,dx + B\,dy$ then a necessary and sufficient condition for df to be exact is

$$\dfrac{\partial A(x, y)}{\partial y} = \dfrac{\partial B(x, y)}{\partial x}.$$

Here $A = -(y^2 + xy)$ and $B = x^2$, and so we calculate

$$\dfrac{\partial(x^2)}{\partial x} = 2x \quad \text{and} \quad \dfrac{\partial(-y^2 - xy)}{\partial y} = -2y - x.$$

These are not equal and so df is *not* an exact differential.

However, for dg, $A = -(y^2 + xy)/(xy^2)$ and $B = x^2/(xy^2)$. Taking the appropriate partial derivatives gives

$$\dfrac{\partial}{\partial x}\left(\dfrac{x^2}{xy^2}\right) = \dfrac{1}{y^2} \quad \text{and} \quad \dfrac{\partial}{\partial y}\left(\dfrac{-y^2 - xy}{xy^2}\right) = 0 + \dfrac{1}{y^2}.$$

These are equal, implying that dg is an exact differential and that the original inexact differential has $1/xy^2$ as its integrating factor.

7.5 The equation $3y = z^3 + 3xz$ defines z implicitly as a function of x and y. Evaluate all three second partial derivatives of z with respect to x and/or y. Verify that z is a solution of

$$x\frac{\partial^2 z}{\partial y^2} + \frac{\partial^2 z}{\partial x^2} = 0.$$

By successive partial differentiations of

$$3y = z^3 + 3xz \qquad (*)$$

and its derivatives with respect to (wrt) x and y, we obtain the following.

Of $(*)$ wrt x
$$0 = 3z^2\frac{\partial z}{\partial x} + 3z + 3x\frac{\partial z}{\partial x},$$

(i)
$$\Rightarrow \quad \frac{\partial z}{\partial x} = -\frac{z}{x + z^2}.$$

Of $(*)$ wrt y
$$3 = 3z^2\frac{\partial z}{\partial y} + 3x\frac{\partial z}{\partial y},$$

(ii)
$$\Rightarrow \quad \frac{\partial z}{\partial y} = \frac{1}{x + z^2}.$$

For the second derivatives:

differentiating (i) wrt x
$$\frac{\partial^2 z}{\partial x^2} = -\frac{(x + z^2)\frac{\partial z}{\partial x} - z\left(1 + 2z\frac{\partial z}{\partial x}\right)}{(x + z^2)^2}$$

$$= \frac{(z^2 - x)\frac{\partial z}{\partial x} + z}{(x + z^2)^2}$$

$$= \frac{(z^2 - x)(-z) + z(x + z^2)}{(x + z^2)^3}, \qquad \text{using (i)},$$

$$= \frac{2xz}{(x + z^2)^3};$$

differentiating (i) wrt y
$$\frac{\partial^2 z}{\partial y\partial x} = -\frac{(x + z^2)\frac{\partial z}{\partial y} - z\,2z\frac{\partial z}{\partial y}}{(x + z^2)^2}$$

$$= \frac{(z^2 - x)\frac{\partial z}{\partial y}}{(x + z^2)^2}$$

$$= \frac{z^2 - x}{(x + z^2)^3}, \qquad \text{using (ii);}$$

differentiating (ii) wrt y

$$\frac{\partial^2 z}{\partial y^2} = \frac{-1}{(x+z^2)^2} 2z \frac{\partial z}{\partial y}$$

$$= \frac{-2z}{(x+z^2)^3}, \qquad \text{using (ii).}$$

We now have that

$$x \frac{\partial^2 z}{\partial y^2} + \frac{\partial^2 z}{\partial x^2} = \frac{-2zx}{(x+z^2)^3} + \frac{2zx}{(x+z^2)^3} = 0,$$

i.e. z is a solution of the given partial differential equation.

7.7 The function $G(t)$ is defined by

$$G(t) = F(x, y) = x^2 + y^2 + 3xy,$$

where $x(t) = at^2$ and $y(t) = 2at$. Use the chain rule to find the values of (x, y) at which $G(t)$ has stationary values as a function of t. Do any of them correspond to the stationary points of $F(x, y)$ as a function of x and y?

Using the chain rule,

$$\frac{dG}{dt} = \frac{\partial F}{\partial x} \frac{dx}{dt} + \frac{\partial F}{\partial y} \frac{dy}{dt}$$

$$= (2x + 3y)2at + (2y + 3x)2a$$

$$= 2at(2at^2 + 6at) + 2a(4at + 3at^2)$$

$$= 2a^2 t(2t^2 + 9t + 4)$$

$$= 2a^2 t(2t + 1)(t + 4).$$

Thus dG/dt has zeros at $t = 0$, $t = -\frac{1}{2}$ and $t = -4$; the corresponding values of (x, y) are $(0, 0)$, $(\frac{1}{4}a, -a)$ and $(16a, -8a)$.

Considered as a function of x and y, $F(x, y)$ has stationary points when

$$\frac{\partial F}{\partial x} = 2x + 3y = 0,$$

$$\frac{\partial F}{\partial y} = 3x + 2y = 0.$$

The *only* solution to this pair of equations is $(x, y) = (0, 0)$, which corresponds to (only) one of the points found previously. This stationary point is a saddle point at the origin and is the only stationary point of $F(x, y)$.

The stationary points of $G(t)$ as a function of t are a maximum of $5a^2/16$ at $(\frac{1}{4}a, -a)$, a minimum of $-64a^2$ at $(16a, -8a)$, and a point of inflection at the origin. The first two are not stationary points of $F(x, y)$ for general values of x and y. They only appear to be so because the parameterisation, which restricts the search to the (one-dimensional) line defined by the parabola $y^2 = 4ax$, does not take into account the values of $F(x, y)$ at points close to, but not on, the line.

7.9 The function $f(x, y)$ satisfies the differential equation

$$y\frac{\partial f}{\partial x} + x\frac{\partial f}{\partial y} = 0.$$

By changing to new variables $u = x^2 - y^2$ and $v = 2xy$, show that f is, in fact, a function of $x^2 - y^2$ only.

In order to use the equations

$$\frac{\partial f}{\partial x_j} = \sum_{i=1}^{n} \frac{\partial f}{\partial u_i}\frac{\partial u_i}{\partial x_j}$$

that govern a change of variables, we need the partial derivatives

$$\frac{\partial u}{\partial x} = 2x, \quad \frac{\partial u}{\partial y} = -2y, \quad \frac{\partial v}{\partial x} = 2y, \quad \frac{\partial v}{\partial y} = 2x.$$

Then, with $f(x, y)$ written as $g(u, v)$,

$$\frac{\partial f}{\partial x} = 2x\frac{\partial g}{\partial u} + 2y\frac{\partial g}{\partial v},$$

$$\frac{\partial f}{\partial y} = -2y\frac{\partial g}{\partial u} + 2x\frac{\partial g}{\partial v}.$$

Thus,

$$y\frac{\partial f}{\partial x} + x\frac{\partial f}{\partial y} = (2xy - 2xy)\frac{\partial g}{\partial u} + 2(y^2 + x^2)\frac{\partial g}{\partial v}.$$

But, from the initial equation, the LHS is equal to zero and the equation reduces to

$$\frac{\partial g}{\partial v} = 0 \quad \Rightarrow \quad g = g(u), \text{ i.e. } f(x, y) = g(x^2 - y^2) \text{ only.}$$

7.11 Find and evaluate the maxima, minima and saddle points of the function

$$f(x, y) = xy(x^2 + y^2 - 1).$$

The required derivatives are given by

$$\frac{\partial f}{\partial x} = 3x^2 y + y^3 - y, \quad \frac{\partial f}{\partial y} = x^3 + 3y^2 x - x,$$

$$\frac{\partial^2 f}{\partial x^2} = 6xy, \quad \frac{\partial^2 f}{\partial x \partial y} = 3x^2 + 3y^2 - 1, \quad \frac{\partial^2 f}{\partial y^2} = 6xy.$$

Any stationary points must satisfy both of the equations

$$\frac{\partial f}{\partial x} = y(3x^2 + y^2 - 1) = 0,$$

$$\frac{\partial f}{\partial y} = x(x^2 + 3y^2 - 1) = 0.$$

If $x = 0$ then $y = 0$ or ± 1. If $y = 0$ then $x = 0$ or ± 1.

Otherwise, adding and subtracting the factors in parentheses gives

$$4(x^2 + y^2) = 2,$$
$$2(x^2 - y^2) = 0.$$

These have the solutions $x = \pm\frac{1}{2}$, $y = \pm\frac{1}{2}$.

Thus the nine stationary points are $(0, 0)$, $(0, \pm 1)$, $(\pm 1, 0)$, $\pm(\frac{1}{2}, \frac{1}{2})$ and $\pm(\frac{1}{2}, -\frac{1}{2})$. The corresponding values for $f(x, y)$ are 0 for the first five, $-\frac{1}{8}$ for the next two and $\frac{1}{8}$ for the final two.

For the first five cases, $\partial^2 f/\partial^2 x = \partial^2 f/\partial^2 y = 0$, whilst $\partial^2 f/\partial x \partial y = -1$ or 2. Since $(-1)^2 > 0 \times 0$ and $2^2 > 0 \times 0$, these points are all saddle points.

At $\pm(\frac{1}{2}, \frac{1}{2})$, $\partial^2 f/\partial^2 x = \partial^2 f/\partial^2 y = \frac{3}{2}$, whilst $\partial^2 f/\partial x \partial y = \frac{1}{2}$. Since $(\frac{1}{2})^2 < \frac{3}{2} \times \frac{3}{2}$, these two points are either maxima or minima (i.e. not saddle points) and the positive signs for $\partial^2 f/\partial^2 x$ and $\partial^2 f/\partial^2 y$ indicate that they are, in fact, minima.

At $\pm(\frac{1}{2}, -\frac{1}{2})$, $\partial^2 f/\partial^2 x = \partial^2 f/\partial^2 y = -\frac{3}{2}$, whilst $\partial^2 f/\partial x \partial y = \frac{1}{2}$. Since $(\frac{1}{2})^2 < -\frac{3}{2} \times -\frac{3}{2}$, these two points are also either maxima or minima; the common negative sign for $\partial^2 f/\partial^2 x$ and $\partial^2 f/\partial^2 y$ indicates that they are maxima.

7.13 Locate the stationary points of the function

$$f(x, y) = (x^2 - 2y^2) \exp[-(x^2 + y^2)/a^2],$$

where a is a non-zero constant.

Sketch the function along the x- and y-axes and hence identify the nature and values of the stationary points.

To find the stationary points, we set each of the two first partial derivatives,

$$\frac{\partial f}{\partial x} = \left[2x - \frac{2x}{a^2}(x^2 - 2y^2)\right] \exp\left(-\frac{x^2 + y^2}{a^2}\right),$$

$$\frac{\partial f}{\partial y} = \left[-4y - \frac{2y}{a^2}(x^2 - 2y^2)\right] \exp\left(-\frac{x^2 + y^2}{a^2}\right),$$

equal to zero:

$$\frac{\partial f}{\partial x} = 0 \quad \Rightarrow \quad x = 0 \text{ or } x^2 - 2y^2 = a^2;$$

$$\frac{\partial f}{\partial y} = 0 \quad \Rightarrow \quad y = 0 \text{ or } x^2 - 2y^2 = -2a^2.$$

Since $a \neq 0$, possible solutions for (x, y) are $(0, 0)$, $(0, \pm a)$ and $(\pm a, 0)$. The corresponding values are $f(0, 0) = 0$, $f(0, \pm a) = -2a^2 e^{-1}$ and $f(\pm a, 0) = a^2 e^{-1}$. These results, taken together with the observation that $|f(x, y)| \rightarrow 0$ as either or both of $|x|$ and $|y| \rightarrow \infty$, show that $f(x, y)$ has maxima at $(\pm a, 0)$, minima at $(0, \pm a)$ and a saddle point at the origin.

Sketches of $f(x, 0)$ and $f(0, y)$, whilst hardly necessary, illustrate rather than confirm these conclusions.

7.15 Find the stationary values of

$$f(x, y) = 4x^2 + 4y^2 + x^4 - 6x^2y^2 + y^4$$

and classify them as maxima, minima or saddle points. Make a rough sketch of the contours of f in the quarter plane $x, y \geq 0$.

The required derivatives are as follows:

$$\frac{\partial f}{\partial x} = 8x + 4x^3 - 12xy^2, \qquad \frac{\partial f}{\partial y} = 8y - 12x^2y + 4y^3,$$

$$\frac{\partial^2 f}{\partial x^2} = 8 + 12x^2 - 12y^2, \qquad \frac{\partial^2 f}{\partial x \partial y} = -24xy, \qquad \frac{\partial^2 f}{\partial y^2} = 8 - 12x^2 + 12y^2.$$

Any stationary points must satisfy both of the equations

$$\frac{\partial f}{\partial x} = 4x(2 + x^2 - 3y^2) = 0,$$

$$\frac{\partial f}{\partial y} = 4y(2 - 3x^2 + y^2) = 0.$$

If $x = 0$ then $4y(2 + y^2) = 0$, implying that $y = 0$ also, since $2 + y^2 = 0$ has no real solutions. Conversely, $y = 0$ implies $x = 0$. Further solutions exist if both expressions in parentheses equal zero; this requires $x^2 = y^2 = 1$.

Thus the stationary points are $(0, 0)$, $(1, 1)$, $(-1, 1)$, $(1, -1)$ and $(-1, -1)$, with corresponding values 0, 4, 4, 4 and 4.

At $(0, 0)$, $\partial^2 f/\partial^2 x = \partial^2 f/\partial^2 y = 8$, whilst $\partial^2 f/\partial x \partial y = 0$. Since $0^2 < 8 \times 8$, this point is a minimum.

In the other four cases, $\partial^2 f/\partial^2 x = \partial^2 f/\partial^2 y = 8$, whilst $\partial^2 f/\partial x \partial y = \pm 24$. Since $(24)^2 > 8 \times 8$, these four points are all saddle points.

It will probably be helpful when sketching the contours (Figure 7.1) to determine the behaviour of $f(x, y)$ along the line $x = y$ and to note the symmetry about it. In particular, note that $f(x, y) = 0$ at both the origin and the point $(\sqrt{2}, \sqrt{2})$.

7.17 A rectangular parallelepiped has all eight vertices on the ellipsoid

$$x^2 + 3y^2 + 3z^2 = 1.$$

Using the symmetry of the parallelepiped about each of the planes $x = 0$, $y = 0$, $z = 0$, write down the surface area of the parallelepiped in terms of the coordinates of the vertex that lies in the octant $x, y, z \geq 0$. Hence find the maximum value of the surface area of such a parallelepiped.

Let S be the surface area and (x, y, z) the coordinates of one of the corners of the parallelepiped with x, y and z all positive. Then we need to maximise $S = 8(xy + yz + zx)$ subject to x, y and z satisfying $x^2 + 3y^2 + 3z^2 = 1$.

Consider

$$f(x, y, z) = 8(xy + yz + zx) + \lambda(x^2 + 3y^2 + 3z^2),$$

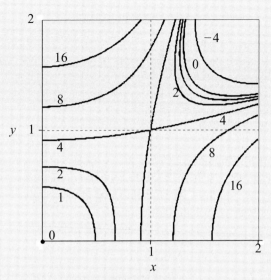

Figure 7.1 The contours found in Problem 7.15.

where λ is a Lagrange undetermined multiplier. Then, setting each of the first partial derivatives separately to zero, we have the simultaneous equations

$$0 = \frac{\partial f}{\partial x} = 8y + 8z + 2\lambda x,$$

$$0 = \frac{\partial f}{\partial y} = 8x + 8z + 6\lambda y,$$

$$0 = \frac{\partial f}{\partial z} = 8x + 8y + 6\lambda z.$$

From symmetry, $y = z$, leading to

$$0 = 16y + 2\lambda x,$$

$$0 = 8x + 8y + 6\lambda y.$$

Thus, rejecting the trivial solution $x = 0$, $y = 0$, we conclude that $\lambda = -8y/x$, leading to $x^2 + xy - 6y^2 = (x - 2y)(x + 3y) = 0$. The only solution to this quadratic equation with x, y and z all positive is $x = 2y = 2z$. Substituting this into the equation of the ellipse gives

$$(2y)^2 + 3y^2 + 3y^2 = 1 \qquad \Rightarrow \qquad y = \frac{1}{\sqrt{10}}.$$

The value of S is then given by

$$S = 8\left(\frac{2}{10} + \frac{1}{10} + \frac{2}{10}\right) = 4.$$

7.19 A barn is to be constructed with a uniform cross-sectional area A throughout its length. The cross-section is to be a rectangle of wall height h (fixed) and width w, surmounted by an isosceles triangular roof that makes an angle θ with the horizontal. The cost of construction is α per unit height of wall and β per unit (slope) length of roof. Show that, irrespective of the values of α and β, to minimise costs w should be chosen to satisfy the equation

$$w^4 = 16A(A - wh),$$

and θ made such that $2 \tan 2\theta = w/h$.

The cost *always* includes $2\alpha h$ for the vertical walls, which can therefore be ignored in the minimisation procedure. The rest of the calculation will be solely concerned with minimising the roof area, and the optimum choices for w and θ will be independent of β, the actual cost per unit length of the roof.

The cost of the roof is $2\beta \times \frac{1}{2} w \sec \theta$, but w and θ are constrained by the requirement that

$$A = wh + \frac{1}{2} w \frac{w}{2} \tan \theta.$$

So we consider $G(w, \theta)$, where

$$G(w, \theta) = \beta w \sec \theta - \lambda(wh + \tfrac{1}{4} w^2 \tan \theta),$$

and the implications of equating its partial derivatives to zero. The first derivative to be set to zero is

$$\frac{\partial G}{\partial \theta} = \beta w \sec \theta \tan \theta - \frac{\lambda}{4} w^2 \sec^2 \theta,$$

$$\Rightarrow \quad 0 = \beta \sin \theta - \tfrac{1}{4} \lambda w,$$

$$\Rightarrow \quad \lambda = \frac{4\beta \sin \theta}{w}.$$

A second equation is provided by differentiation with respect to w and yields

$$\frac{\partial G}{\partial w} = \beta \sec \theta - \lambda h - \tfrac{1}{2} \lambda w \tan \theta.$$

Setting $\partial G / \partial w = 0$, multiplying through by $\cos \theta$ and substituting for λ, we obtain

$$\beta - 2\beta \sin^2 \theta = \frac{4\beta \sin \theta\, h \cos \theta}{w},$$

$$w \cos 2\theta = 2h \sin 2\theta,$$

$$\tan 2\theta = \frac{w}{2h}.$$

This is the second result quoted.

The overall area constraint can be written

$$\tan \theta = \frac{4(A - wh)}{w^2}.$$

From these two results and the double-angle formula $\tan 2\phi = 2\tan\phi/(1 - \tan^2\phi)$, it follows that

$$\frac{w}{2h} = \tan 2\theta$$

$$= \frac{\dfrac{8(A - wh)}{w^2}}{1 - \dfrac{16(A - wh)^2}{w^4}},$$

$$16wh(A - wh) = w^4 - 16(A - wh)^2,$$

$$w^4 = 16A(A - wh).$$

This is the first quoted result, and we note that, as expected, both optimum values are independent of β.

7.21 Find the area of the region covered by points on the lines

$$\frac{x}{a} + \frac{y}{b} = 1,$$

where the sum of any line's intercepts on the coordinate axes is fixed and equal to c.

The equation of a typical line with intercept a on the x-axis is

$$f(x, y, a) = \frac{x}{a} + \frac{y}{c - a} - 1 = 0.$$

To find the envelope of the lines we set $\partial f/\partial a = 0$. This gives

$$\frac{\partial f}{\partial a} = -\frac{x}{a^2} + \frac{y}{(c - a)^2} = 0.$$

Hence,

$$(c - a)\sqrt{x} = a\sqrt{y},$$

$$a = \frac{c\sqrt{x}}{\sqrt{x} + \sqrt{y}}.$$

Substituting this value into $f(x, y, a) = 0$ gives the equation of the envelope as

$$\frac{x(\sqrt{x} + \sqrt{y})}{c\sqrt{x}} + \frac{y}{c - \dfrac{c\sqrt{x}}{\sqrt{x} + \sqrt{y}}} = 1,$$

$$\sqrt{x}(\sqrt{x} + \sqrt{y}) + \sqrt{y}(\sqrt{x} + \sqrt{y}) = c,$$

$$\sqrt{x} + \sqrt{y} = \sqrt{c}.$$

This is a curve (not a straight line) whose end-points are $(c, 0)$ on the x-axis and $(0, c)$ on the y-axis. All points on lines with the given property lie below this envelope curve (except for one point on each line, which lies on the curve). Consequently, the area covered by the points is that bounded by the envelope and the two axes. It has the value

$$\int_0^c y\,dx = \int_0^c (\sqrt{c} - \sqrt{x})^2\,dx$$

$$= \int_0^c (c - 2\sqrt{c}\,\sqrt{x} + x)\,dx$$

$$= c^2 - \tfrac{4}{3}\sqrt{c}\,c^{3/2} + \tfrac{1}{2}c^2 = \tfrac{1}{6}c^2.$$

7.23 A water feature contains a spray head at water level at the centre of a round basin. The head is in the form of a small hemisphere perforated by many evenly distributed small holes, through which water spurts out at the same speed, v_0, in all directions.

(a) What is the shape of the 'water bell' so formed?
(b) What must be the minimum diameter of the bowl if no water is to be lost?

The system has cylindrical symmetry and so we work with cylindrical polar coordinates ρ and z.

For a jet of water emerging from the spray head at an angle θ to the vertical, the equations of motion are

$$z = v_0 \cos\theta\, t - \tfrac{1}{2}gt^2,$$

$$\rho = v_0 \sin\theta\, t.$$

Eliminating the time, t, we have

$$z = \frac{\rho\, v_0 \cos\theta}{v_0 \sin\theta} - \frac{1}{2}g\frac{\rho^2}{v_0^2 \sin^2\theta},$$

$$\Rightarrow \quad 0 = z - \rho \cot\theta + \frac{g\rho^2}{2v_0^2}\operatorname{cosec}^2\theta;$$

i.e. with $\cot\theta$ written as α, the trajectory of this jet is given by

$$f(\rho, z, \alpha) = z - \rho\alpha + \frac{g\rho^2}{2v_0^2}(1 + \alpha^2) = 0.$$

To find the envelope of all these trajectories as θ (and hence α) is varied, we set $\partial f/\partial\alpha$ equal to zero:

$$0 = \frac{\partial f}{\partial\alpha} = 0 - \rho + \frac{2\alpha g\rho^2}{2v_0^2},$$

$$\Rightarrow \quad \alpha = \frac{v_0^2}{g\rho}.$$

Hence, the equation of the envelope, and thus of the water bell, is

$$g(\rho, z) = z - \frac{v_0^2}{g} + \frac{g\rho^2}{2v_0^2}\left(1 + \frac{v_0^4}{g^2\rho^2}\right) = 0,$$

$$\Rightarrow \quad z = \frac{v_0^2}{2g} - \frac{g\rho^2}{2v_0^2}.$$

(a) This is the equation of a parabola whose apex is at $z = v_0^2/2g$, $\rho = 0$. It follows that the water bell has the shape of an inverted paraboloid of revolution.

(b) When $z = 0$, ρ has the value v_0^2/g, and hence the minimum value needed for the diameter of the bowl is given by $2\rho = 2v_0^2/g$.

7.25 By considering the differential

$$dG = d(U + PV - ST),$$

where G is the Gibbs free energy, P the pressure, V the volume, S the entropy and T the temperature of a system, and given further that U, the internal energy, satisfies

$$dU = T\,dS - P\,dV,$$

derive a Maxwell relation connecting $(\partial V/\partial T)_P$ and $(\partial S/\partial P)_T$.

Given that $dU = T\,dS - P\,dV$, we have that

$$dG = d(U + PV - ST)$$
$$= dU + P\,dV + V\,dP - S\,dT - T\,dS$$
$$= V\,dP - S\,dT.$$

Hence,

$$\left(\frac{\partial G}{\partial P}\right)_T = V \quad \text{and} \quad \left(\frac{\partial G}{\partial T}\right)_P = -S.$$

It follows that

$$\left(\frac{\partial V}{\partial T}\right)_P = \frac{\partial^2 G}{\partial T \partial P} = \frac{\partial^2 G}{\partial P \partial T} = -\left(\frac{\partial S}{\partial P}\right)_T.$$

This is the required Maxwell thermodynamic relation.

7.27 As implied in Problem 7.25 on the thermodynamics of a simple gas, the quantity $dS = T^{-1}(dU + P\,dV)$ is an exact differential. Use this to prove that

$$\left(\frac{\partial U}{\partial V}\right)_T = T\left(\frac{\partial P}{\partial T}\right)_V - P.$$

In the van der Waals model of a gas, P obeys the equation

$$P = \frac{RT}{V - b} - \frac{a}{V^2},$$

where R, a and b are constants. Further, in the limit $V \to \infty$, the form of U becomes $U = cT$, where c is another constant. Find the complete expression for $U(V, T)$.

Writing the total differentials in $dS = T^{-1}(dU + P\,dV)$ in terms of partial derivatives with respect to V and T gives

$$T\left(\frac{\partial S}{\partial V}\right)_T dV + T\left(\frac{\partial S}{\partial T}\right)_V dT = \left(\frac{\partial U}{\partial V}\right)_T dV + \left(\frac{\partial U}{\partial T}\right)_V dT + P\,dV,$$

from which it follows that

$$T\left(\frac{\partial S}{\partial V}\right)_T = \left(\frac{\partial U}{\partial V}\right)_T + P \quad (*) \quad \text{and} \quad T\left(\frac{\partial S}{\partial T}\right)_V = \left(\frac{\partial U}{\partial T}\right)_V.$$

Differentiating the first of these with respect to T and the second with respect to V, and then combining the two equations so obtained, gives

$$\left(\frac{\partial S}{\partial V}\right)_T + T\frac{\partial^2 S}{\partial T \partial V} = \frac{\partial^2 U}{\partial T \partial V} + \left(\frac{\partial P}{\partial T}\right)_V,$$

$$T\frac{\partial^2 S}{\partial V \partial T} = \frac{\partial^2 U}{\partial V \partial T}.$$

Taken together, these imply that $\left(\frac{\partial S}{\partial V}\right)_T = \left(\frac{\partial P}{\partial T}\right)_V.$

The equation $(*)$ can now be written in the required form:

$$\left(\frac{\partial U}{\partial V}\right)_T = T\left(\frac{\partial P}{\partial T}\right)_V - P.$$

For the van der Waals model gas,

$$P = \frac{RT}{V - b} - \frac{a}{V^2},$$

and we can substitute for P in the previous result to give

$$\left(\frac{\partial U}{\partial V}\right)_T = T\left(\frac{R}{V - b}\right) - \left(\frac{RT}{V - b} - \frac{a}{V^2}\right) = \frac{a}{V^2},$$

which integrates to

$$U(V, T) = -\frac{a}{V} + f(T).$$

Since $U \to cT$ as $V \to \infty$ for all T, the unknown function, $f(T)$, must be simply $f(T) = cT$. Thus, the full expression for $U(V, T)$ is

$$U(V, T) = cT - \frac{a}{V}.$$

We note that, in the limit $V \to \infty$, van der Waals' equation becomes $PV = RT$ and thus recognise c as the specific heat at constant volume of a perfect gas.

7.29 By finding dI/dy, evaluate the integral

$$I(y) = \int_0^\infty \frac{e^{-xy} \sin x}{x}\, dx.$$

Hence show that

$$J = \int_0^\infty \frac{\sin x}{x}\, dx = \frac{\pi}{2}.$$

Since the integral is over positive values of x, its convergence requires that $y \geq 0$. We first express the $\sin x$ factor as a complex exponential:

$$I(y) = \int_0^\infty \frac{e^{-xy} \sin x}{x} \, dx$$

$$= \operatorname{Im} \int_0^\infty \frac{e^{-xy+ix}}{x} \, dx.$$

And now differentiate under the integral sign:

$$\frac{dI}{dy} = \operatorname{Im} \int_0^\infty \frac{(-x)e^{-xy+ix}}{x} \, dx$$

$$= \operatorname{Im} \left[\frac{-e^{-xy+ix}}{-y+i} \right]_0^\infty$$

$$= \operatorname{Im} \left(\frac{1}{-y+i} \right)$$

$$= -\frac{1}{1+y^2}.$$

This differential equation expresses how the integral varies as a function of y. But, as we can see immediately that for $y = \infty$ the integral must be zero, we can find its value for non-infinite y by integrating the differential equation:

$$I(y) - I(\infty) = \int_\infty^y \frac{-1}{1+y^2} \, dy = -\tan^{-1} y + \tan^{-1} \infty = \frac{\pi}{2} - \tan^{-1} y.$$

In the limit $y \to 0$ this becomes

$$J = \int_0^\infty \frac{\sin x}{x} \, dx = I(0) = \frac{\pi}{2} - 0 = \frac{\pi}{2}.$$

7.31 The function $f(x)$ is differentiable and $f(0) = 0$. A second function $g(y)$ is defined by

$$g(y) = \int_0^y \frac{f(x) \, dx}{\sqrt{y-x}}.$$

Prove that

$$\frac{dg}{dy} = \int_0^y \frac{df}{dx} \frac{dx}{\sqrt{y-x}}.$$

For the case $f(x) = x^n$, prove that

$$\frac{d^n g}{dy^n} = 2(n!)\sqrt{y}.$$

Integrating the definition of $g(y)$ by parts:

$$g(y) = \int_0^y \frac{f(x)\,dx}{\sqrt{y-x}}$$

$$= \left[-2f(x)\sqrt{y-x} \right]_0^y + \int_0^y 2\frac{df}{dx}\sqrt{y-x}\,dx$$

$$= 2\int_0^y \frac{df}{dx}\sqrt{y-x},$$

where we have used $f(0) = 0$ in setting the definite integral to zero.

Now, differentiating $g(y)$ with respect to both its upper limit and its integrand, we obtain

$$\frac{dg}{dy} = 2\frac{df}{dx}\sqrt{y-y} + 2\int_0^y \frac{1}{2}\frac{df}{dx}\frac{1}{\sqrt{y-x}} = \int_0^y \frac{df}{dx}\frac{1}{\sqrt{y-x}}.$$

This result, showing that the construction of the derivative of g from the derivative of f is the same as that of g from f, applies to any function that satisfies $f(0) = 0$ and so applies to x^n and all of its derivatives. It follows that

$$\frac{d^n g}{dy^n} = \int_0^y \frac{d^n f}{dx^n}\frac{1}{\sqrt{y-x}}\,dx$$

$$= \int_0^y \frac{n!}{\sqrt{y-x}}\,dx$$

$$= \left[\frac{n!(-1)\sqrt{y-x}}{\frac{1}{2}} \right]_0^y$$

$$= 2(n!)\sqrt{y}.$$

7.33   If

$$I(\alpha) = \int_0^1 \frac{x^\alpha - 1}{\ln x}\,dx, \qquad \alpha > -1,$$

what is the value of $I(0)$? Show that

$$\frac{d}{d\alpha}x^\alpha = x^\alpha \ln x$$

and deduce that

$$\frac{d}{d\alpha}I(\alpha) = \frac{1}{\alpha+1}.$$

Hence prove that $I(\alpha) = \ln(1+\alpha)$.

Since the integrand is singular at $x = 1$, we need to define $I(0)$ as a limit:

$$I(0) = \lim_{y\to 1}\int_0^y \frac{x^0 - 1}{\ln x}\,dx = \lim_{y\to 1}\int_0^y 0\,dx = \lim_{y\to 1} 0 = 0,$$

i.e. $I(0) = 0$.

With $z = x^\alpha$, we have

$$\ln z = \alpha \ln x \quad \Rightarrow \quad \frac{1}{z}\frac{dz}{d\alpha} = \ln x$$

$$\Rightarrow \quad \frac{dz}{d\alpha} = z \ln x \quad \Rightarrow \quad \frac{d}{d\alpha} x^\alpha = x^\alpha \ln x.$$

The derivative of $I(\alpha)$ is then

$$\frac{dI}{d\alpha} = \int_0^1 \frac{1}{\ln x} x^\alpha \ln x \, dx$$

$$= \left[\frac{x^{\alpha+1}}{\alpha+1} \right]_0^1$$

$$= \frac{1}{\alpha+1}.$$

Finally, integration gives

$$I(\alpha) - I(0) = \int_0^\alpha \frac{d\beta}{\beta+1},$$

$$I(\alpha) - 0 = \ln(1+\alpha).$$

To obtain this final line we have used our first result that $I(0) = 0$.

7.35 The function $G(t, \xi)$ is defined for $0 \le t \le \pi$ by

$$G(t, \xi) = \begin{cases} -\cos t \sin \xi & \text{for } \xi \le t, \\ -\sin t \cos \xi & \text{for } \xi > t. \end{cases}$$

Show that the function $x(t)$ defined by

$$x(t) = \int_0^\pi G(t, \xi) f(\xi) \, d\xi$$

satisfies the equation

$$\frac{d^2 x}{dt^2} + x = f(t),$$

where $f(t)$ can be *any* arbitrary (continuous) function. Show further that $x(0) = [dx/dt]_{t=\pi} = 0$, again for any $f(t)$, but that the *value* of $x(\pi)$ does depend upon the form of $f(t)$.

[The function $G(t, \xi)$ is an example of a Green's function, an important concept in the solution of differential equations.]

The explicit integral expression for $x(t)$ is

$$x(t) = \int_0^\pi G(t, \xi) f(\xi) \, d\xi$$

$$= - \int_0^t \cos t \sin \xi \, f(\xi) \, d\xi - \int_t^\pi \sin t \cos \xi \, f(\xi) \, d\xi.$$

We now form its first two derivatives using Leibnitz's rule:

$$\frac{dx}{dt} = -\cos t[\sin t\, f(t)] + \sin t \int_0^t \sin\xi\, f(\xi)\,d\xi$$

$$+ \sin t[\cos t\, f(t)] - \cos t \int_t^{\pi} \cos\xi\, f(\xi)\,d\xi$$

$$= \sin t \int_0^t \sin\xi\, f(\xi)\,d\xi - \cos t \int_t^{\pi} \cos\xi\, f(\xi)\,d\xi.$$

$$\frac{d^2x}{dt^2} = \cos t \int_0^t \sin\xi\, f(\xi)\,d\xi + \sin t[\sin t\, f(t)]$$

$$+ \sin t \int_t^{\pi} \cos\xi\, f(\xi)\,d\xi + \cos t[\cos t\, f(t)]$$

$$= -x(t) + f(t)(\sin^2 t + \cos^2 t).$$

This shows that

$$\frac{d^2x}{dt^2} + x = f(t)$$

for *any* continuous function $f(x)$.

When $t = 0$ the first integral in the expression for $x(t)$ has zero range and the second is multiplied by $\sin 0$; consequently $x(0) = 0$.

When $t = \pi$ the second integral in the expression for dx/dt has zero range and the first is multiplied by $\sin\pi$; consequently $[dx/dt]_{t=\pi} = 0$.

However, when $t = \pi$, although the second integral in the expression for $x(t)$ is multiplied by $\sin\pi$ and contributes nothing, the first integral is not zero in general and its value *will* depend upon the form of $f(t)$.

8 Multiple integrals

8.1 Identify the curved wedge bounded by the surfaces $y^2 = 4ax$, $x + z = a$ and $z = 0$, and hence calculate its volume V.

As will readily be seen from a rough sketch, the wedge consists of that part of a parabolic cylinder, parallel to the z-axis, that is cut off by two planes, one parallel to the y-axis and the other the coordinate plane $z = 0$.

For the first stage of the multiple integration, the volume can be divided equally easily into 'vertical columns' or into horizontal strips parallel to the y-axis. Thus there are two equivalent and equally obvious ways of proceeding.

Either

$$V = \int_0^a dx \int_{-\sqrt{4ax}}^{\sqrt{4ax}} dy \int_0^{a-x} dz$$

$$= \int_0^a 2\sqrt{4ax}(a - x)\, dx$$

$$= 4\sqrt{a}\left[\tfrac{2}{3}ax^{3/2} - \tfrac{2}{5}x^{5/2}\right]_0^a = \tfrac{16}{15}a^3;$$

or

$$V = \int_0^a dz \int_0^{a-z} dx \int_{-\sqrt{4ax}}^{\sqrt{4ax}} dy$$

$$= \int_0^a dz \int_0^{a-z} 2\sqrt{4ax}\, dx$$

$$= 4\sqrt{a} \int \tfrac{2}{3}(a - z)^{3/2}\, dz$$

$$= \frac{8\sqrt{a}}{3}\left[-\tfrac{2}{5}(a - z)^{5/2}\right]_0^a = \tfrac{16}{15}a^3.$$

As is to be expected, the calculated volume is the same, irrespective of the order in which the integrations are carried out.

8.3 Find the volume integral of x^2y over the tetrahedral volume bounded by the planes $x = 0$, $y = 0$, $z = 0$ and $x + y + z = 1$.

The bounding surfaces of the integration volume are symmetric in x, y and z and, on these grounds, there is nothing to choose between the various possible orders of integration.

However, the integrand does not contain z and so there is some advantage in carrying out the z-integration first. Its value can simply be set equal to the length of the z-interval and the dimension of the integral will have been reduced by one 'at a stroke'.

$$
\begin{aligned}
I &= \int_0^1 dx \int_0^{1-x} dy \int_0^{1-x-y} x^2 y \, dz \\
&= \int_0^1 dx \int_0^{1-x} x^2 y (1 - x - y) \, dy \\
&= \int_0^1 \left[x^2 (1 - x) \frac{(1 - x)^2}{2} - x^2 \frac{(1 - x)^3}{3} \right] dx \\
&= \frac{1}{6} \int_0^1 x^2 (1 - 3x + 3x^2 - x^3) \, dx \\
&= \frac{1}{6} \left(\frac{1}{3} - \frac{3}{4} + \frac{3}{5} - \frac{1}{6} \right) \\
&= \frac{1}{6} \frac{20 - 45 + 36 - 10}{60} = \frac{1}{360}.
\end{aligned}
$$

8.5 Calculate the volume of an ellipsoid as follows:

(a) Prove that the area of the ellipse

$$
\frac{x^2}{a^2} + \frac{y^2}{b^2} = 1
$$

is πab.

(b) Use this result to obtain an expression for the volume of a slice of thickness dz of the ellipsoid

$$
\frac{x^2}{a^2} + \frac{y^2}{b^2} + \frac{z^2}{c^2} = 1.
$$

Hence show that the volume of the ellipsoid is $4\pi abc/3$.

(a) Dividing the ellipse into thin strips parallel to the y-axis, we may write its area as

$$
\text{area} = 2 \int_{-a}^{a} y \, dx = 2 \int_{-a}^{a} b \sqrt{1 - \left(\frac{x}{a} \right)^2} \, dx.
$$

Set $x = a \cos \phi$ with $dx = -a \sin \phi \, d\phi$. Then

$$
\text{area} = 2b \int_{\pi}^{0} \sin \phi (-a \sin \phi) \, d\phi = 2ab \int_0^{\pi} \sin^2 \phi \, d\phi = 2ab \frac{\pi}{2} = \pi ab.
$$

(b) Consider slices of the ellipsoid, of thickness dz, taken perpendicular to the z-axis. Each is an ellipse whose bounding curve is given by the equation

$$\frac{x^2}{a^2} + \frac{y^2}{b^2} = 1 - \frac{z^2}{c^2}$$

and is thus a scaled-down version of the ellipse considered in part (a) with semi-axes $a[1 - (z/c)^2]^{1/2}$ and $b[1 - (z/c)^2]^{1/2}$. Its area is therefore $\pi a[1 - (z/c)^2]^{1/2} b[1 - (z/c)^2]^{1/2}$ and its volume dV is this multiplied by dz. Thus, the total volume V of the ellipsoid is given by

$$\int_{-c}^{c} \pi ab \left(1 - \frac{z^2}{c^2}\right) dz = \pi ab \left[z - \frac{1}{3}\frac{z^3}{c^2}\right]_{-c}^{c} = \frac{4\pi abc}{3}.$$

8.7 In quantum mechanics the electron in a hydrogen atom in some particular state is described by a wavefunction Ψ, which is such that $|\Psi|^2 \, dV$ is the probability of finding the electron in the infinitesimal volume dV. In spherical polar coordinates $\Psi = \Psi(r, \theta, \phi)$ and $dV = r^2 \sin\theta \, dr \, d\theta \, d\phi$. Two such states are described by

$$\Psi_1 = \left(\frac{1}{4\pi}\right)^{1/2} \left(\frac{1}{a_0}\right)^{3/2} 2e^{-r/a_0},$$

$$\Psi_2 = -\left(\frac{3}{8\pi}\right)^{1/2} \sin\theta \, e^{i\phi} \left(\frac{1}{2a_0}\right)^{3/2} \frac{re^{-r/2a_0}}{a_0\sqrt{3}}.$$

(a) Show that each Ψ_i is normalised, i.e. the integral over all space $\int |\Psi|^2 \, dV$ is equal to unity – physically, this means that the electron must be somewhere.

(b) The (so-called) dipole matrix element between the states 1 and 2 is given by the integral

$$p_x = \int \Psi_1^* q r \sin\theta \cos\phi \, \Psi_2 \, dV,$$

where q is the charge on the electron. Prove that p_x has the value $-2^7 q a_0/3^5$.

(a) We need to show that the volume integral of $|\Psi_i|^2$ is equal to unity, and begin by noting that, since ϕ is not explicitly mentioned, or appears only in the form $e^{i\phi}$, the ϕ integration of $|\Psi|^2$ yields a factor of 2π in each case. For Ψ_1 we have

$$\int |\Psi_1|^2 \, dV = \int |\Psi_1|^2 r^2 \sin\theta \, d\theta \, d\phi \, dr$$

$$= \frac{1}{4\pi} \frac{4}{a_0^3} 2\pi \int_0^\infty r^2 e^{-2r/a_0} \, dr \int_0^\pi \sin\theta \, d\theta$$

$$= \frac{2}{a_0^3} \int_0^\infty 2r^2 e^{-2r/a_0} \, dr$$

$$= \frac{4}{a_0^3} \frac{a_0}{2} 2 \frac{a_0}{2} 1 \frac{a_0}{2} = 1.$$

The last line has been obtained using repeated integration by parts.

For Ψ_2, the corresponding calculation is

$$\int |\Psi_2|^2 \, dV = \int |\Psi_2|^2 r^2 \sin\theta \, d\theta \, d\phi \, dr$$

$$= \frac{2\pi}{64\pi a_0^5} \int_0^\infty r^4 e^{-r/a_0} \, dr \int_0^\pi \sin^3\theta \, d\theta$$

$$= \frac{1}{32 a_0^5} \int_0^\infty r^4 e^{-r/a_0} \, dr \int_0^\pi (1 - \cos^2\theta) \sin\theta \, d\theta$$

$$= \frac{1}{32 a_0^5} \, 4! \, a_0^5 \left(2 - \frac{2}{3} \right) = 1.$$

Again, the r-integral was calculated using integration by parts. In summary, both functions are correctly normalised.

(b) The dipole matrix element has important physical properties, but for the purposes of this problem it is simply an integral to be evaluated according to a formula, as follows:

$$p_x = \int \Psi_1^* q r \sin\theta \cos\phi \, \Psi_2 \, r^2 \sin\theta \, d\theta \, d\phi \, dr$$

$$= \frac{-q}{8\pi a_0^4} \int_0^\pi \sin^3\theta \, d\theta \int_0^{2\pi} \cos\phi(\cos\phi + i \sin\phi) \, d\phi \int_0^\infty r^4 e^{-3r/2a_0} \, dr$$

$$= -\frac{q}{8\pi a_0^4} \left(2 - \frac{2}{3} \right) (\pi + i0) \, 4! \left(\frac{2a_0}{3} \right)^5$$

$$= -\frac{2^7}{3^5} q a_0.$$

8.9 A certain torus has a circular vertical cross-section of radius a centred on a horizontal circle of radius $c \, (> a)$.

(a) Find the volume V and surface area A of the torus, and show that they can be written as

$$V = \frac{\pi^2}{4}(r_o^2 - r_i^2)(r_o - r_i), \qquad A = \pi^2(r_o^2 - r_i^2),$$

where r_o and r_i are, respectively, the outer and inner radii of the torus.

(b) Show that a vertical circular cylinder of radius c, coaxial with the torus, divides A in the ratio

$$\pi c + 2a \; : \; \pi c - 2a.$$

(a) The inner and outer radii of the torus are $r_i = c - a$ and $r_o = c + a$, from which it follows that $r_o^2 - r_i^2 = 4ac$ and that $r_o - r_i = 2a$.

The torus is generated by sweeping the centre of a circle of radius a, area πa^2 and circumference $2\pi a$ around a circle of radius c. Therefore, by Pappus's first theorem, the volume of the torus is given by

$$V = \pi a^2 \times 2\pi c = 2\pi^2 a^2 c = \frac{\pi^2}{4}(r_o^2 - r_i^2)(r_o - r_i),$$

whilst, by his second theorem, its surface area is

$$A = 2\pi a \times 2\pi c = 4\pi^2 ac = \pi^2 (r_o^2 - r_i^2).$$

(b) The vertical cylinder divides the perimeter of a cross-section of the torus into two equal parts. The distance from the cylinder of the centroid of either half is given by

$$\bar{x} = \frac{\int x \, ds}{\int ds} = \frac{\int_{-\pi/2}^{\pi/2} a \cos\phi \, a \, d\phi}{\int_{-\pi/2}^{\pi/2} a \, d\phi} = \frac{2a}{\pi}.$$

It therefore follows from Pappus's second theorem that

$$A_o = \pi a \times 2\pi \left(c + \frac{2a}{\pi} \right) \quad \text{and} \quad A_i = \pi a \times 2\pi \left(c - \frac{2a}{\pi} \right),$$

leading to the stated result.

8.11 In some applications in mechanics the moment of inertia of a body about a single point (as opposed to about an axis) is needed. The moment of inertia, I, about the origin of a uniform solid body of density ρ is given by the volume integral

$$I = \int_V (x^2 + y^2 + z^2)\rho \, dV.$$

Show that the moment of inertia of a right circular cylinder of radius a, length $2b$ and mass M about its centre is given by

$$M \left(\frac{a^2}{2} + \frac{b^2}{3} \right).$$

Since the cylinder is easily described in cylindrical polar coordinates (ρ, ϕ, z), we convert the calculation to one using those coordinates and denote the density by ρ_0 to avoid confusion:

$$I = \int_V (x^2 + y^2 + z^2)\rho_0 \, dV$$

$$= \rho_0 \int_V (\rho^2 + z^2)\rho \, d\rho \, d\phi \, dz$$

$$= \rho_0 \int_0^{2\pi} d\phi \int_0^a \rho \, d\rho \int_{-b}^b (\rho^2 + z^2) \, dz$$

$$= 2\pi\rho_0 \int_0^a \rho \left(2b\rho^2 + \frac{2b^3}{3} \right) d\rho$$

$$= 2\pi\rho_0 \left(2b\frac{a^4}{4} + \frac{2b^3}{3}\frac{a^2}{2} \right).$$

Now $M = \pi a^2 \times 2b \times \rho_0$, and so the moment of inertia about the origin can be expressed as

$$I = M \left(\frac{a^2}{2} + \frac{b^2}{3} \right).$$

8.13 In spherical polar coordinates r, θ, ϕ the element of volume for a body that is symmetrical about the polar axis is $dV = 2\pi r^2 \sin\theta \, dr \, d\theta$, whilst its element of surface area is $2\pi r \sin\theta[(dr)^2 + r^2(d\theta)^2]^{1/2}$. A particular surface is defined by $r = 2a \cos\theta$, where a is a constant and $0 \le \theta \le \pi/2$. Find its total surface area and the volume it encloses, and hence identify the surface.

With the surface of the body defined by $r = 2a \cos\theta$, for calculating its total volume the radial integration variable r' lies in the range $0 \le r' \le 2a \cos\theta$. Hence

$$V = \int_0^{\pi/2} 2\pi \sin\theta \, d\theta \int_0^{2a \cos\theta} r'^2 \, dr'$$

$$= 2\pi \int_0^{\pi/2} \sin\theta \, \frac{(2a \cos\theta)^3}{3} \, d\theta$$

$$= \frac{16\pi a^3}{3} \int_0^{\pi/2} \cos^3\theta \sin\theta \, d\theta$$

$$= \frac{16\pi a^3}{3} \left[-\frac{\cos^4\theta}{4} \right]_0^{\pi/2}$$

$$= \tfrac{4}{3}\pi a^3.$$

The additional strip of surface area resulting from a change from θ to $\theta + d\theta$ is $2\pi r \sin\theta \, d\ell$, where $d\ell$ is the length of the generating curve that lies in this infinitesimal range of θ. This is given by

$$(d\ell)^2 = (dr)^2 + (r \, d\theta)^2$$

$$= (-2a \sin\theta \, d\theta)^2 + (2a \cos\theta \, d\theta)^2$$

$$= 4a^2 (d\theta)^2$$

The integral becomes one-dimensional with

$$S = 2\pi \int_0^{\pi/2} 2a \cos\theta \sin\theta \, 2a \, d\theta$$

$$= 8\pi a^2 \left[\frac{\sin^2\theta}{2} \right]_0^{\pi/2}$$

$$= 4\pi a^2.$$

With a volume of $\tfrac{4}{3}\pi a^3$ and a surface area of $4\pi a^2$, the surface is probably that of a sphere of radius a, with the origin at the 'lowest' point of the sphere. This conclusion is confirmed by the fact that the triangle formed by the two ends of the vertical diameter of the sphere and any point on its surface is a right-angled triangle in which $r/2a = \cos\theta$.

8.15 By transforming to cylindrical polar coordinates, evaluate the integral

$$I = \int\int\int \ln(x^2 + y^2)\, dx\, dy\, dz$$

over the interior of the conical region $x^2 + y^2 \le z^2, 0 \le z \le 1$.

The volume element $dx\, dy\, dz$ becomes $\rho\, d\rho\, d\phi\, dz$ in cylindrical polar coordinates and the integrand contains a factor $\rho \ln \rho^2 = 2\rho \ln \rho$. This is dealt with using integration by parts and the integral becomes

$$I = \int\int\int 2\rho \ln \rho\, d\rho\, d\phi\, dz \quad \text{over} \quad \rho \le z,\ 0 \le z \le 1,$$

$$= 2 \int_0^{2\pi} d\phi \int_0^1 dz \int_0^z \rho \ln \rho\, d\rho$$

$$= 2 \cdot 2\pi \int_0^1 \left(\left[\frac{\rho^2 \ln \rho}{2} \right]_0^z - \int_0^z \frac{1}{\rho} \frac{\rho^2}{2}\, d\rho \right) dz$$

$$= 4\pi \int_0^1 \left(\frac{1}{2} z^2 \ln z - \frac{1}{4} z^2 \right) dz$$

$$= 2\pi \left(\left[\frac{z^3 \ln z}{3} \right]_0^1 - \int_0^1 \frac{1}{z} \frac{z^3}{3}\, dz \right) - \pi \left[\frac{z^3}{3} \right]_0^1$$

$$= 2\pi \left(0 - \left[\frac{z^3}{9} \right]_0^1 \right) - \frac{\pi}{3}$$

$$= -\frac{2\pi}{9} - \frac{\pi}{3} = -\frac{5\pi}{9}.$$

Although the integrand contains no explicit minus signs, a negative value for the integral is to be expected, since $1 \ge z^2 \ge x^2 + y^2$ and $\ln(x^2 + y^2)$ is therefore negative.

8.17 By making two successive simple changes of variables, evaluate

$$I = \int\int\int x^2\, dx\, dy\, dz$$

over the ellipsoidal region

$$\frac{x^2}{a^2} + \frac{y^2}{b^2} + \frac{z^2}{c^2} \le 1.$$

We start by making a scaling change aimed at producing an integration volume that has more amenable properties than an ellipsoid, namely a sphere. To do this, set $\xi = x/a$,

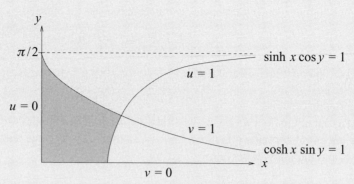

Figure 8.1 The integration area for Problem 8.19.

$\eta = y/b$ and $\zeta = z/c$; the integral then becomes

$$I = \int\int\int a^2\xi^2 \, a \, d\xi \, b \, d\eta \, c \, d\zeta \quad \text{over} \quad \xi^2 + \eta^2 + \zeta^2 \le 1$$

$$= a^3 bc \int\int\int \xi^2 \, d\xi \, d\eta \, d\zeta.$$

With the integration volume now a sphere it is sensible to change to spherical polar variables: $\xi = r\cos\theta$, $\eta = r\sin\theta\cos\phi$ and $\zeta = r\sin\theta\sin\phi$, with volume element $d\xi \, d\eta \, d\zeta = r^2 \sin\theta \, dr \, d\theta \, d\phi$. Note that we have chosen to orientate the polar axis along the old x-axis, rather than along the more conventional z-axis.

$$I = a^3 bc \int_0^{2\pi} d\phi \int_0^{\pi} \cos^2\theta \sin\theta \, d\theta \int_0^1 r^4 \, dr$$

$$= a^3 bc \, 2\pi \, \frac{2}{3} \frac{1}{5}$$

$$= \tfrac{4}{15} \pi a^3 bc.$$

8.19 Sketch that part of the region $0 \le x$, $0 \le y \le \pi/2$ which is bounded by the curves $x = 0$, $y = 0$, $\sinh x \cos y = 1$ and $\cosh x \sin y = 1$. By making a suitable change of variables, evaluate the integral

$$I = \int\int (\sinh^2 x + \cos^2 y) \sinh 2x \sin 2y \, dx \, dy$$

over the bounded subregion.

The integration area is shaded in Figure 8.1. We are guided in making a choice of new variables by the equations defining the 'awkward' parts of the subregion's boundary curve. Ideally, the new variables should each be constant along one or more of the curves making up the boundary. This consideration leads us to make a change to new variables, $u = \sinh x \cos y$ and $v = \cosh x \sin y$. We then find the following.

(i) The boundary $y = 0$ becomes $v = 0$.

(ii) The boundary $x = 0$ becomes $u = 0$.

(iii) The boundary $\sinh x \cos y = 1$ becomes $u = 1$.

(iv) The boundary $\cosh x \sin y = 1$ becomes $v = 1$.

With this choice for the change, all four parts of the boundary can be characterised as being lines along which one of the coordinates is constant.

The Jacobian relating $dx\,dy$ to $du\,dv$, i.e. $du\,dv = \dfrac{\partial(u,\,v)}{\partial(x,\,y)}\,dx\,dy$, is

$$\frac{\partial(u,\,v)}{\partial(x,\,y)} = \frac{\partial u}{\partial x}\frac{\partial v}{\partial y} - \frac{\partial u}{\partial y}\frac{\partial v}{\partial x}$$

$$= (\cosh x \cos y)(\cosh x \cos y) - (-\sinh x \sin y)(\sinh x \sin y)$$

$$= (\sinh^2 x + 1)\cos^2 y + \sinh^2 x \sin^2 y$$

$$= \sinh^2 x + \cos^2 y.$$

The Jacobian required for the change of variables in the current case is the inverse of this.

Making the change of variables, and recalling that $\sin 2z = 2 \sin z \cos z$, and similarly for $\sinh 2z$, gives

$$I = \int\int (\sinh^2 x + \cos^2 y)\sinh 2x \sin 2y\,dx\,dy$$

$$= \int_0^1\int_0^1 (\sinh^2 x + \cos^2 y)(4uv)\,\frac{du\,dv}{\sinh^2 x + \cos^2 y}$$

$$= 4\int_0^1 u\,du \int_0^1 v\,dv$$

$$= 4\left[\frac{u^2}{2}\right]_0^1 \left[\frac{v^2}{2}\right]_0^1 = 1.$$

This is the simple answer to a superficially difficult integral!

8.21 As stated in Section 7, the first law of thermodynamics can be expressed as

$$dU = T\,dS - P\,dV.$$

By calculating and equating $\partial^2 U/\partial Y \partial X$ and $\partial^2 U/\partial X \partial Y$, where X and Y are an unspecified pair of variables (drawn from P, V, T and S), prove that

$$\frac{\partial(S,\,T)}{\partial(X,\,Y)} = \frac{\partial(V,\,P)}{\partial(X,\,Y)}.$$

Using the properties of Jacobians, deduce that

$$\frac{\partial(S,\,T)}{\partial(V,\,P)} = 1.$$

Starting from

$$dU = T\,dS - P\,dV,$$

the partial derivatives of U with respect to X and Y are

$$\frac{\partial U}{\partial X} = T\frac{\partial S}{\partial X} - P\frac{\partial V}{\partial X} \quad \text{and} \quad \frac{\partial U}{\partial Y} = T\frac{\partial S}{\partial Y} - P\frac{\partial V}{\partial Y}.$$

We next differentiate these two expressions to obtain two (equal) second derivatives. Note that, since X and Y can be any pair drawn from P, V, T and S, we must differentiate all four terms on the RHS as products, giving rise to two terms each. The resulting equations are

$$\frac{\partial^2 U}{\partial Y\partial X} = T\frac{\partial^2 S}{\partial Y\partial X} + \frac{\partial T}{\partial Y}\frac{\partial S}{\partial X} - P\frac{\partial^2 V}{\partial Y\partial X} - \frac{\partial P}{\partial Y}\frac{\partial V}{\partial X},$$

$$\frac{\partial^2 U}{\partial X\partial Y} = T\frac{\partial^2 S}{\partial X\partial Y} + \frac{\partial T}{\partial X}\frac{\partial S}{\partial Y} - P\frac{\partial^2 V}{\partial X\partial Y} - \frac{\partial P}{\partial X}\frac{\partial V}{\partial Y}.$$

Equating the two expressions, and then cancelling the terms that appear on both side of the equality, yields

$$\frac{\partial T}{\partial Y}\frac{\partial S}{\partial X} - \frac{\partial P}{\partial Y}\frac{\partial V}{\partial X} = \frac{\partial T}{\partial X}\frac{\partial S}{\partial Y} - \frac{\partial P}{\partial X}\frac{\partial V}{\partial Y},$$

$$\Rightarrow \quad \frac{\partial T}{\partial Y}\frac{\partial S}{\partial X} - \frac{\partial T}{\partial X}\frac{\partial S}{\partial Y} = \frac{\partial P}{\partial Y}\frac{\partial V}{\partial X} - \frac{\partial P}{\partial X}\frac{\partial V}{\partial Y},$$

$$\Rightarrow \quad \frac{\partial(S, T)}{\partial(X, Y)} = \frac{\partial(V, P)}{\partial(X, Y)}.$$

Now, using this result and the properties of Jacobians ($J_{pr} = J_{pq}J_{qr}$ and $J_{pq} = [J_{qp}]^{-1}$), we can write

$$\frac{\partial(S, T)}{\partial(V, P)} = \frac{\partial(S, T)}{\partial(X, Y)}\frac{\partial(X, Y)}{\partial(V, P)}$$

$$= \frac{\partial(S, T)}{\partial(X, Y)}\left[\frac{\partial(V, P)}{\partial(X, Y)}\right]^{-1}$$

$$= \frac{\partial(S, T)}{\partial(X, Y)}\left[\frac{\partial(S, T)}{\partial(X, Y)}\right]^{-1}$$

$$= 1.$$

9 Vector algebra

9.1 Which of the following statements about general vectors **a**, **b** and **c** are true?

(a) $\mathbf{c} \cdot (\mathbf{a} \times \mathbf{b}) = (\mathbf{b} \times \mathbf{a}) \cdot \mathbf{c}$;
(b) $\mathbf{a} \times (\mathbf{b} \times \mathbf{c}) = (\mathbf{a} \times \mathbf{b}) \times \mathbf{c}$;
(c) $\mathbf{a} \times (\mathbf{b} \times \mathbf{c}) = (\mathbf{a} \cdot \mathbf{c})\mathbf{b} - (\mathbf{a} \cdot \mathbf{b})\mathbf{c}$;
(d) $\mathbf{d} = \lambda \mathbf{a} + \mu \mathbf{b}$ implies $(\mathbf{a} \times \mathbf{b}) \cdot \mathbf{d} = 0$;
(e) $\mathbf{a} \times \mathbf{c} = \mathbf{b} \times \mathbf{c}$ implies $\mathbf{c} \cdot \mathbf{a} - \mathbf{c} \cdot \mathbf{b} = c\,|\mathbf{a} - \mathbf{b}|$;
(f) $(\mathbf{a} \times \mathbf{b}) \times (\mathbf{c} \times \mathbf{b}) = \mathbf{b}[\,\mathbf{b} \cdot (\mathbf{c} \times \mathbf{a})]$.

All of the tests below are made using combinations of the common properties of the various types of vector products and justifications for individual steps are therefore not given. If the properties used are not recognised, they can be found in and learned from almost any standard textbook.

(a) $\mathbf{c} \cdot (\mathbf{a} \times \mathbf{b}) = -\mathbf{c} \cdot (\mathbf{b} \times \mathbf{a}) = -(\mathbf{b} \times \mathbf{a}) \cdot \mathbf{c} \neq (\mathbf{b} \times \mathbf{a}) \cdot \mathbf{c}$.
(b) $\mathbf{a} \times (\mathbf{b} \times \mathbf{c}) = \mathbf{b}(\mathbf{a} \cdot \mathbf{c}) - \mathbf{c}(\mathbf{a} \cdot \mathbf{b}) \neq \mathbf{b}(\mathbf{a} \cdot \mathbf{c}) - \mathbf{a}(\mathbf{b} \cdot \mathbf{c}) = (\mathbf{a} \times \mathbf{b}) \times \mathbf{c}$.
(c) $\mathbf{a} \times (\mathbf{b} \times \mathbf{c}) = (\mathbf{a} \cdot \mathbf{c})\mathbf{b} - (\mathbf{a} \cdot \mathbf{b})\mathbf{c}$, a standard result.
(d) $(\mathbf{a} \times \mathbf{b}) \cdot \mathbf{d} = (\mathbf{a} \times \mathbf{b}) \cdot (\lambda \mathbf{a} + \mu \mathbf{b}) = \lambda(\mathbf{a} \times \mathbf{b}) \cdot \mathbf{a} + \mu(\mathbf{a} \times \mathbf{b}) \cdot \mathbf{b} = \lambda 0 + \mu 0 = 0$.
(e) $\mathbf{a} \times \mathbf{c} = \mathbf{b} \times \mathbf{c} \Rightarrow (\mathbf{a} - \mathbf{b}) \times \mathbf{c} = 0 \Rightarrow \mathbf{a} - \mathbf{b} \parallel \mathbf{c} \Rightarrow (\mathbf{a} - \mathbf{b}) \cdot \mathbf{c} = c\,|\mathbf{a} - \mathbf{b}| \Rightarrow$
  $\mathbf{c} \cdot \mathbf{a} - \mathbf{c} \cdot \mathbf{b} = c\,|\mathbf{a} - \mathbf{b}|$.
(f) $(\mathbf{a} \times \mathbf{b}) \times (\mathbf{c} \times \mathbf{b}) = \mathbf{b}[\,\mathbf{a} \cdot (\mathbf{c} \times \mathbf{b})] - \mathbf{a}[\,\mathbf{b} \cdot (\mathbf{c} \times \mathbf{b})] = \mathbf{b}[\,\mathbf{a} \cdot (\mathbf{c} \times \mathbf{b})] - 0 =$
  $\mathbf{b}[\,\mathbf{b} \cdot (\mathbf{a} \times \mathbf{c})] = -\mathbf{b}[\,\mathbf{b} \cdot (\mathbf{c} \times \mathbf{a})] \neq \mathbf{b}[\,\mathbf{b} \cdot (\mathbf{c} \times \mathbf{a})]$.

Thus only (c), (d) and (e) are true.

9.3 Identify the following surfaces:

(a) $|\mathbf{r}| = k$; (b) $\mathbf{r} \cdot \mathbf{u} = l$; (c) $\mathbf{r} \cdot \mathbf{u} = m|\mathbf{r}|$ for $-1 \leq m \leq +1$;
(d) $|\mathbf{r} - (\mathbf{r} \cdot \mathbf{u})\mathbf{u}| = n$.

Here k, l, m and n are fixed scalars and **u** is a fixed unit vector.

(a) All points on the surface are a distance k from the origin. The surface is therefore a sphere of radius k centred on the origin.

(b) This is the standard vector equation of a plane whose normal is in the direction **u** and whose distance from the origin is l.

(c) This is the surface generated by all vectors that make an angle $\cos^{-1} m$ with the fixed unit vector **u**. The surface is therefore the cone of semi-angle $\cos^{-1} m$ that has the direction of **u** as its axis and the origin as its vertex.

(d) Since $(\mathbf{r} \cdot \mathbf{u})\mathbf{u}$ is the component of $\mathbf{r}$ that is parallel to $\mathbf{u}$, $\mathbf{r} - (\mathbf{r} \cdot \mathbf{u})\mathbf{u}$ is the component perpendicular to $\mathbf{u}$. As this latter component is constant for all points on the surface, the surface must be a circular cylinder of radius n that has its axis parallel to $\mathbf{u}$.

9.5 A, B, C and D are the four corners, in order, of one face of a cube of side 2 units. The opposite face has corners E, F, G and H, with AE, BF, CG and DH as parallel edges of the cube. The centre O of the cube is taken as the origin and the x-, y- and z- axes are parallel to AD, AE and AB, respectively. Find the following:

(a) the angle between the face diagonal AF and the body diagonal AG;
(b) the equation of the plane through B that is parallel to the plane CGE;
(c) the perpendicular distance from the centre J of the face $BCGF$ to the plane OCG;
(d) the volume of the tetrahedron $JOCG$.

(a) Unit vectors in the directions of the two diagonals have components

$$\mathbf{f} - \mathbf{a} = \frac{(0, 2, 2)}{\sqrt{8}} \quad \text{and} \quad \mathbf{g} - \mathbf{a} = \frac{(2, 2, 2)}{\sqrt{12}}.$$

Taking the scalar product of these two unit vectors gives the angle between them as

$$\theta = \cos^{-1} \frac{0 + 4 + 4}{\sqrt{96}} = \cos^{-1} \sqrt{\frac{2}{3}}.$$

(b) The direction of a normal $\mathbf{n}$ to the plane CGE is in the direction of the cross product of any two non-parallel vectors that lie in the plane. These can be taken as those from C to G and from C to E:

$$(\mathbf{g} - \mathbf{c}) \times (\mathbf{e} - \mathbf{c}) = (0, 2, 0) \times (-2, 2, -2) = (-4, 0, 4).$$

The equation of the plane is therefore of the form

$$c = \mathbf{n} \cdot \mathbf{r} = -4x + 0y + 4z = -4x + 4z.$$

Since it passes through $\mathbf{b} = (-1, -1, 1)$, the value of c must be 8 and the equation of the plane is $z - x = 2$.

(c) The direction of a normal $\mathbf{n}$ to the plane OCG is given by

$$\mathbf{c} \times \mathbf{g} = (1, -1, 1) \times (1, 1, 1) = (-2, 0, 2).$$

The equation of the plane is therefore of the form

$$c = \mathbf{n} \cdot \mathbf{r} = -2x + 0y + 2z = -2x + 2z.$$

Since it passes through the origin, the value of c must be zero and the equation of the plane written in the form $\hat{\mathbf{n}} \cdot \mathbf{r} = p$ is

$$-\frac{x}{\sqrt{2}} + \frac{z}{\sqrt{2}} = 0.$$

The distance of J from this plane is $\hat{\mathbf{n}} \cdot \mathbf{j}$, where $\mathbf{j} = (0, 0, 1)$. The distance is thus $-0 + (1/\sqrt{2}) = 1/\sqrt{2}$.

(d) The volume of the tetrahedron is $\frac{1}{3}$(base area × height perpendicular to the base). The area of triangle OCG is $\frac{1}{2}|\mathbf{c} \times \mathbf{g}|$ and the perpendicular height of the tetrahedron is the component of $\mathbf{j}$ in the direction of $\mathbf{c} \times \mathbf{g}$. Thus the volume is

$$V = \frac{1}{3}\left|\frac{1}{2}(\mathbf{c} \times \mathbf{g}) \cdot \mathbf{j}\right| = \frac{1}{6}|(-2, 0, 2) \cdot (0, 0, 1)| = \frac{1}{3}.$$

9.7 The edges OP, OQ and OR of a tetrahedron $OPQR$ are vectors $\mathbf{p}$, $\mathbf{q}$ and $\mathbf{r}$, respectively, where $\mathbf{p} = 2\mathbf{i} + 4\mathbf{j}$, $\mathbf{q} = 2\mathbf{i} - \mathbf{j} + 3\mathbf{k}$ and $\mathbf{r} = 4\mathbf{i} - 2\mathbf{j} + 5\mathbf{k}$. Show that OP is perpendicular to the plane containing OQR. Express the volume of the tetrahedron in terms of $\mathbf{p}$, $\mathbf{q}$ and $\mathbf{r}$ and hence calculate the volume.

The plane containing OQR has a normal in the direction $\mathbf{q} \times \mathbf{r} = (2, -1, 3) \times (4, -2, 5) = (1, 2, 0)$. This is parallel to $\mathbf{p}$ since $\mathbf{q} \times \mathbf{r} = \frac{1}{2}\mathbf{p}$. The volume of the tetrahedron is therefore one-third times $\frac{1}{2}|\mathbf{q} \times \mathbf{r}|$ times $|\mathbf{p}|$, i.e. $\frac{1}{6}|(1, 2, 0)|\sqrt{20} = \frac{5}{3}$.

9.9 Prove Lagrange's identity, i.e.

$$(\mathbf{a} \times \mathbf{b}) \cdot (\mathbf{c} \times \mathbf{d}) = (\mathbf{a} \cdot \mathbf{c})(\mathbf{b} \cdot \mathbf{d}) - (\mathbf{a} \cdot \mathbf{d})(\mathbf{b} \cdot \mathbf{c}).$$

We treat the expression on the LHS as the triple scalar product of the three vectors $\mathbf{a} \times \mathbf{b}$, $\mathbf{c}$ and $\mathbf{d}$ and use the cyclic properties of triple scalar products:

$$(\mathbf{a} \times \mathbf{b}) \cdot (\mathbf{c} \times \mathbf{d}) = \mathbf{d} \cdot [(\mathbf{a} \times \mathbf{b}) \times \mathbf{c}]$$
$$= \mathbf{d} \cdot [(\mathbf{a} \cdot \mathbf{c})\mathbf{b} - (\mathbf{b} \cdot \mathbf{c})\mathbf{a}]$$
$$= (\mathbf{a} \cdot \mathbf{c})(\mathbf{d} \cdot \mathbf{b}) - (\mathbf{b} \cdot \mathbf{c})(\mathbf{d} \cdot \mathbf{a}).$$

In going from the first to the second line we used the standard result

$$(\mathbf{a} \times \mathbf{b}) \times \mathbf{c} = (\mathbf{a} \cdot \mathbf{c})\mathbf{b} - (\mathbf{b} \cdot \mathbf{c})\mathbf{a}$$

to replace $(\mathbf{a} \times \mathbf{b}) \times \mathbf{c}$. This result, if not known, can be proved by writing it out in component form as follows.

Consider only the x-component of each side of the equation. The corresponding results for other components can be obtained by cyclic permutation of x, y and z.

$$\mathbf{a} \times \mathbf{b} = (a_y b_z - a_z b_y, \ a_z b_x - a_x b_z, \ a_x b_y - a_y b_x)$$
$$[(\mathbf{a} \times \mathbf{b}) \times \mathbf{c}]_x = (a_z b_x - a_x b_z)c_z - (a_x b_y - a_y b_x)c_y$$
$$= b_x(a_z c_z + a_y c_y) - a_x(b_z c_z + b_y c_y)$$
$$= b_x(a_z c_z + a_y c_y + a_x c_x) - a_x(b_x c_x + b_z c_z + b_y c_y)$$
$$= [(\mathbf{a} \cdot \mathbf{c})\mathbf{b} - (\mathbf{b} \cdot \mathbf{c})\mathbf{a}]_x.$$

To obtain the penultimate line we both added and subtracted $a_x b_x c_x$ on the RHS. This establishes the result for the x-component and hence for all three components.

9.11 Show that the points $(1, 0, 1)$, $(1, 1, 0)$ and $(1, -3, 4)$ lie on a straight line. Give the equation of the line in the form

$$\mathbf{r} = \mathbf{a} + \lambda \mathbf{b}.$$

To show that the points lie on a line, we need to show that their position vectors are linearly dependent. That this is so follows from noting that

$$(1, -3, 4) = 4(1, 0, 1) - 3(1, 1, 0).$$

This can also be written

$$(1, -3, 4) = (1, 0, 1) + 3[(1, 0, 1) - (1, 1, 0)] = (1, 0, 1) + 3(0, -1, 1).$$

The equation of the line is therefore

$$\mathbf{r} = \mathbf{a} + \lambda(-\mathbf{j} + \mathbf{k}),$$

where $\mathbf{a}$ is the vector position of *any* point on the line, e.g. $\mathbf{i} + \mathbf{k}$ or $\mathbf{i} + \mathbf{j}$ or $\mathbf{i} - 3\mathbf{j} + 4\mathbf{k}$ or many others. Of course, choosing different points for $\mathbf{a}$ will entail using different values of λ to describe the same point $\mathbf{r}$ on the line. For example,

$$(1, -5, 6) = (1, 0, 1) + 5(0, -1, 1)$$
$$\text{or} = (1, 1, 0) + 6(0, -1, 1)$$
$$\text{or} = (1, -3, 4) + 2(0, -1, 1).$$

9.13 Two planes have non-parallel unit normals $\hat{\mathbf{n}}$ and $\hat{\mathbf{m}}$ and their closest distances from the origin are λ and μ, respectively. Find the vector equation of their line of intersection in the form $\mathbf{r} = \nu\mathbf{p} + \mathbf{a}$.

The equations of the two planes are

$$\hat{\mathbf{n}} \cdot \mathbf{r} = \lambda \qquad \text{and} \qquad \hat{\mathbf{m}} \cdot \mathbf{r} = \mu.$$

The line of intersection lies in both planes and is thus perpendicular to both normals; it therefore has direction $\mathbf{p} = \hat{\mathbf{n}} \times \hat{\mathbf{m}}$. Consequently, the equation of the line takes the form $\mathbf{r} = \nu\mathbf{p} + \mathbf{a}$, where $\mathbf{a}$ is any one point lying on it. One such point is the one in which the line meets the plane containing $\hat{\mathbf{n}}$ and $\hat{\mathbf{m}}$; we take this point as $\mathbf{a}$. Since $\mathbf{a}$ also lies in both of the original planes, we must have

$$\hat{\mathbf{n}} \cdot \mathbf{a} = \lambda \quad \text{and} \quad \hat{\mathbf{m}} \cdot \mathbf{a} = \mu.$$

If we now write $\mathbf{a} = x\,\hat{\mathbf{n}} + y\,\hat{\mathbf{m}}$, these two conditions become

$$\lambda = \hat{\mathbf{n}} \cdot \mathbf{a} = x + y(\hat{\mathbf{n}} \cdot \hat{\mathbf{m}}),$$
$$\mu = \hat{\mathbf{m}} \cdot \mathbf{a} = x(\hat{\mathbf{n}} \cdot \hat{\mathbf{m}}) + y.$$

It then follows that

$$x = \frac{\lambda - \mu(\hat{\mathbf{n}} \cdot \hat{\mathbf{m}})}{1 - (\hat{\mathbf{n}} \cdot \hat{\mathbf{m}})^2} \quad \text{and} \quad y = \frac{\mu - \lambda(\hat{\mathbf{n}} \cdot \hat{\mathbf{m}})}{1 - (\hat{\mathbf{n}} \cdot \hat{\mathbf{m}})^2},$$

thus determining $\mathbf{a}$. Both $\mathbf{p}$ and $\mathbf{a}$ are therefore determined in terms of λ, μ, $\hat{\mathbf{n}}$ and $\hat{\mathbf{m}}$, and so consequently is the line of intersection of the planes.

9.15 Let O, A, B and C be four points with position vectors $\mathbf{0}$, $\mathbf{a}$, $\mathbf{b}$ and $\mathbf{c}$, and denote by $\mathbf{g} = \lambda\mathbf{a} + \mu\mathbf{b} + \nu\mathbf{c}$ the position of the centre of the sphere on which they all lie.

(a) Prove that λ, μ and ν simultaneously satisfy

$$(\mathbf{a} \cdot \mathbf{a})\lambda + (\mathbf{a} \cdot \mathbf{b})\mu + (\mathbf{a} \cdot \mathbf{c})\nu = \tfrac{1}{2}a^2$$

and two other similar equations.

(b) By making a change of origin, find the centre and radius of the sphere on which the points $\mathbf{p} = 3\mathbf{i} + \mathbf{j} - 2\mathbf{k}$, $\mathbf{q} = 4\mathbf{i} + 3\mathbf{j} - 3\mathbf{k}$, $\mathbf{r} = 7\mathbf{i} - 3\mathbf{k}$ and $\mathbf{s} = 6\mathbf{i} + \mathbf{j} - \mathbf{k}$ all lie.

(a) Each of the points O, A, B and C is the same distance from the centre G of the sphere. In particular, $OG = OA$, i.e.

$$|\mathbf{g} - \mathbf{0}|^2 = |\mathbf{a} - \mathbf{g}|^2,$$
$$g^2 = a^2 - 2\mathbf{a} \cdot \mathbf{g} + g^2,$$
$$\mathbf{a} \cdot \mathbf{g} = \tfrac{1}{2}a^2,$$
$$\mathbf{a} \cdot (\lambda\mathbf{a} + \mu\mathbf{b} + \nu\mathbf{c}) = \tfrac{1}{2}a^2,$$
$$(\mathbf{a} \cdot \mathbf{a})\lambda + (\mathbf{a} \cdot \mathbf{b})\mu + (\mathbf{a} \cdot \mathbf{c})\nu = \tfrac{1}{2}a^2.$$

Two similar equations can be obtained from $OG = OB$ and $OG = OC$.

(b) To use the previous result we make P, say, the origin of a new coordinate system in which

$$\mathbf{p}' = \mathbf{p} - \mathbf{p} = (0, 0, 0),$$
$$\mathbf{q}' = \mathbf{q} - \mathbf{p} = (1, 2, -1),$$
$$\mathbf{r}' = \mathbf{r} - \mathbf{p} = (4, -1, -1),$$
$$\mathbf{s}' = \mathbf{s} - \mathbf{p} = (3, 0, 1).$$

The centre, G, of the sphere on which P, Q, R and S lie is then given by

$$\mathbf{g}' = \lambda\mathbf{q}' + \mu\mathbf{r}' + \nu\mathbf{s}',$$

where

$$(\mathbf{q}' \cdot \mathbf{q}')\lambda + (\mathbf{q}' \cdot \mathbf{r}')\mu + (\mathbf{q}' \cdot \mathbf{s}')\nu = \tfrac{1}{2}\mathbf{q}' \cdot \mathbf{q}',$$
$$(\mathbf{r}' \cdot \mathbf{q}')\lambda + (\mathbf{r}' \cdot \mathbf{r}')\mu + (\mathbf{r}' \cdot \mathbf{s}')\nu = \tfrac{1}{2}\mathbf{r}' \cdot \mathbf{r}',$$
$$(\mathbf{s}' \cdot \mathbf{q}')\lambda + (\mathbf{s}' \cdot \mathbf{r}')\mu + (\mathbf{s}' \cdot \mathbf{s}')\nu = \tfrac{1}{2}\mathbf{s}' \cdot \mathbf{s}',$$

i.e.
$$6\lambda + 3\mu + 2\nu = 3,$$
$$3\lambda + 18\mu + 11\nu = 9,$$
$$2\lambda + 11\mu + 10\nu = 5.$$

These equations have the solution

$$\lambda = \frac{5}{18}, \qquad \mu = \frac{5}{9}, \qquad \nu = -\frac{1}{6}.$$

Thus, the centre of the sphere can be calculated as

$$\mathbf{g}' = \frac{5}{18}(1, 2, -1) + \frac{5}{9}(4, -1, -1) - \frac{1}{6}(3, 0, 1) = (2, 0, -1).$$

Its radius is therefore $|G'O'| = |\mathbf{g}'| = \sqrt{5}$ and its centre in the original coordinate system is at $\mathbf{g}' + \mathbf{p} = (5, 1, -3)$.

9.17 Using vector methods:

(a) Show that the line of intersection of the planes $x + 2y + 3z = 0$ and $3x + 2y + z = 0$ is equally inclined to the x- and z-axes and makes an angle $\cos^{-1}(-2/\sqrt{6})$ with the y-axis.
(b) Find the perpendicular distance between one corner of a unit cube and the major diagonal not passing through it.

(a) The origin O is clearly in both planes. A second such point can be found by setting $z = 1$, say, and solving the pair of simultaneous equations to give $x = 1$ and $y = -2$, i.e. $(1, -2, 1)$ is in both planes. The direction cosines of the line of intersection, OP, are therefore

$$\left(\frac{1}{\sqrt{6}}, -\frac{2}{\sqrt{6}}, \frac{1}{\sqrt{6}} \right),$$

i.e. the line is equally inclined to the x- and z-axes and makes an angle $\cos^{-1}(-2/\sqrt{6})$ with the y-axis.

The same conclusion can be reached by reasoning as follows. The line of intersection of the two planes must be orthogonal to the normal of either plane. Therefore, it is in the direction of the cross product of the two normals and is given by

$$(1, 2, 3) \times (3, 2, 1) = (-4, 8, -4) = -4\sqrt{6}\left(\frac{1}{\sqrt{6}}, -\frac{2}{\sqrt{6}}, \frac{1}{\sqrt{6}} \right).$$

(b) We first note that all three major diagonals not passing through a corner come equally close to it. Taking the corner to be at the origin and the diagonal to be the one that passes through $(0, 1, 1)$ [and $(1, 0, 0)$], the equation of the diagonal is

$$(x, y, z) = (0, 1, 1) + \frac{\lambda}{\sqrt{3}}(1, -1, -1).$$

Using the result that the distance d of the point $\mathbf{p}$ from the line $\mathbf{r} = \mathbf{a} + \lambda \hat{\mathbf{b}}$ is given by

$$d = |(\mathbf{p} - \mathbf{a}) \times \hat{\mathbf{b}}|,$$

the distance of $(0, 0, 0)$ from the line of the diagonal is

$$\left| [(0, 0, 0) - (0, 1, 1)] \times \frac{1}{\sqrt{3}}(1, -1, -1) \right| = \frac{1}{\sqrt{3}}|(0, -1, 1)| = \sqrt{\frac{2}{3}}.$$

[*Note*: the point of closest approach divides the diagonal in the ratio $1 : 2$.]

9.19 The vectors $\mathbf{a}$, $\mathbf{b}$ and $\mathbf{c}$ are not coplanar. Verify that the expressions

$$\mathbf{a}' = \frac{\mathbf{b} \times \mathbf{c}}{[\mathbf{a}, \mathbf{b}, \mathbf{c}]}, \quad \mathbf{b}' = \frac{\mathbf{c} \times \mathbf{a}}{[\mathbf{a}, \mathbf{b}, \mathbf{c}]}, \quad \mathbf{c}' = \frac{\mathbf{a} \times \mathbf{b}}{[\mathbf{a}, \mathbf{b}, \mathbf{c}]}$$

define a set of reciprocal vectors $\mathbf{a}'$, $\mathbf{b}'$ and $\mathbf{c}'$ with the following properties:

(a) $\mathbf{a}' \cdot \mathbf{a} = \mathbf{b}' \cdot \mathbf{b} = \mathbf{c}' \cdot \mathbf{c} = 1$;
(b) $\mathbf{a}' \cdot \mathbf{b} = \mathbf{a}' \cdot \mathbf{c} = \mathbf{b}' \cdot \mathbf{a}$ etc. $= 0$;
(c) $[\mathbf{a}', \mathbf{b}', \mathbf{c}'] = 1/[\mathbf{a}, \mathbf{b}, \mathbf{c}]$;
(d) $\mathbf{a} = (\mathbf{b}' \times \mathbf{c}')/[\mathbf{a}', \mathbf{b}', \mathbf{c}']$.

Direct substitutions and the expansion formula for a vector triple product [in the form given in Equation (9.38)] enable the verifications to be made as follows. We make repeated use of the general result $(\mathbf{p} \times \mathbf{q}) \cdot \mathbf{p} = 0 = (\mathbf{p} \times \mathbf{q}) \cdot \mathbf{q}$.

(a) $\mathbf{a}' \cdot \mathbf{a} = \dfrac{(\mathbf{b} \times \mathbf{c}) \cdot \mathbf{a}}{[\mathbf{a}, \mathbf{b}, \mathbf{c}]} = 1$. Similarly for $\mathbf{b}' \cdot \mathbf{b}$ and $\mathbf{c}' \cdot \mathbf{c}$.

(b) $\mathbf{a}' \cdot \mathbf{b} = \dfrac{(\mathbf{b} \times \mathbf{c}) \cdot \mathbf{b}}{[\mathbf{a}, \mathbf{b}, \mathbf{c}]} = 0$. Similarly for $\mathbf{a}' \cdot \mathbf{c}$, $\mathbf{b}' \cdot \mathbf{a}$ etc.

(c) $[\mathbf{a}', \mathbf{b}', \mathbf{c}'] = \dfrac{\mathbf{a}' \cdot \{(\mathbf{c} \times \mathbf{a}) \times (\mathbf{a} \times \mathbf{b})\}}{[\mathbf{a}, \mathbf{b}, \mathbf{c}]^2}$

$$= \frac{\mathbf{a}' \cdot \{[\mathbf{b} \cdot (\mathbf{c} \times \mathbf{a})]\mathbf{a} - [\mathbf{a} \cdot (\mathbf{c} \times \mathbf{a})]\mathbf{b}\}}{[\mathbf{a}, \mathbf{b}, \mathbf{c}]^2}$$

$$= \frac{1[\mathbf{b}, \mathbf{c}, \mathbf{a}] - 0(\mathbf{a}' \cdot \mathbf{b})}{[\mathbf{a}, \mathbf{b}, \mathbf{c}]^2}, \quad \text{using results (a) and (b),}$$

$$= \frac{1}{[\mathbf{a}, \mathbf{b}, \mathbf{c}]}.$$

(d) $\dfrac{\mathbf{b}' \times \mathbf{c}'}{[\mathbf{a}', \mathbf{b}', \mathbf{c}']} = \dfrac{[\mathbf{b}, \mathbf{c}, \mathbf{a}]\mathbf{a} - 0\mathbf{b}}{[\mathbf{a}, \mathbf{b}, \mathbf{c}]^2 [\mathbf{a}', \mathbf{b}', \mathbf{c}']}, \qquad \text{as in part (c),}$

$$= \mathbf{a}, \qquad\qquad\qquad \text{from result (c).}$$

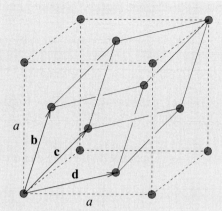

Figure 9.1 A face-centred cubic crystal.

9.21 In a crystal with a face-centred cubic structure, the basic cell can be taken as a cube of edge a with its centre at the origin of coordinates and its edges parallel to the Cartesian coordinate axes; atoms are sited at the eight corners and at the centre of each face. However, other basic cells are possible. One is the rhomboid shown in Figure 9.1, which has the three vectors $\mathbf{b}$, $\mathbf{c}$ and $\mathbf{d}$ as edges.

(a) Show that the volume of the rhomboid is one-quarter that of the cube.

(b) Show that the angles between pairs of edges of the rhomboid are $60°$ and that the corresponding angles between pairs of edges of the rhomboid defined by the reciprocal vectors to $\mathbf{b}$, $\mathbf{c}$, $\mathbf{d}$ are each $109.5°$. (This rhomboid can be used as the basic cell of a body-centred cubic structure, more easily visualised as a cube with an atom at each corner and one at its centre.)

(c) In order to use the Bragg formula, $2d \sin\theta = n\lambda$, for the scattering of X-rays by a crystal, it is necessary to know the perpendicular distance d between successive planes of atoms; for a given crystal structure, d has a particular value for each set of planes considered. For the face-centred cubic structure find the distance between successive planes with normals in the $\mathbf{k}$, $\mathbf{i}+\mathbf{j}$ and $\mathbf{i}+\mathbf{j}+\mathbf{k}$ directions.

(a) From the figure it is easy to see that the edges of the rhomboid are the vectors $\mathbf{b} = \frac{1}{2}a(0, 1, 1)$, $\mathbf{c} = \frac{1}{2}a(1, 0, 1)$ and $\mathbf{d} = \frac{1}{2}a(1, 1, 0)$. The volume V of the rhomboid is therefore given by

$$V = |\,[\,\mathbf{b},\, \mathbf{c},\, \mathbf{d}\,]\,|$$

$$= |\mathbf{b} \cdot (\mathbf{c} \times \mathbf{d})|$$

$$= \tfrac{1}{8}a^3\,|(0, 1, 1) \cdot (-1, 1, 1)|$$

$$= \tfrac{1}{4}a^3,$$

i.e. one-quarter that of the cube.

(b) To find the angle between two edges of the rhomboid we calculate the scalar product of two unit vectors, one along each edge; its value is $1 \times 1 \times \cos \phi$, where ϕ is the angle between the edges. Unit vectors along the edges of the rhomboid are

$$\hat{\mathbf{b}} = \frac{1}{\sqrt{2}}(0, 1, 1), \quad \hat{\mathbf{c}} = \frac{1}{\sqrt{2}}(1, 0, 1), \quad \hat{\mathbf{d}} = \frac{1}{\sqrt{2}}(1, 1, 0).$$

The scalar product of *any* pair of these particular vectors has the value $\frac{1}{2}$, e.g.

$$\hat{\mathbf{b}} \cdot \hat{\mathbf{c}} = \tfrac{1}{2}(0 + 0 + 1) = \tfrac{1}{2}.$$

Thus the angle between any pair of edges is $\cos^{-1}(\frac{1}{2}) = 60°$.

The reciprocal vectors are, for example,

$$\mathbf{b}' = \frac{\mathbf{c} \times \mathbf{d}}{[\mathbf{b}, \mathbf{c}, \mathbf{d}]} = \frac{a^2}{4} \frac{(-1, 1, 1)}{(a^3/4)} = \frac{1}{a}(-1, 1, 1) = \frac{1}{a}(-\mathbf{i} + \mathbf{j} + \mathbf{k}),$$

where in the second equality we have used the result of part (a). Similarly, or by cyclic permutation, $\mathbf{c}' = a^{-1}(\mathbf{i} - \mathbf{j} + \mathbf{k})$ and $\mathbf{d}' = a^{-1}(\mathbf{i} + \mathbf{j} - \mathbf{k})$.

The angle between any pair of reciprocal vectors has the value $109.5°$, e.g.

$$\theta = \cos^{-1}\left(\frac{\mathbf{b}' \cdot \mathbf{c}'}{|\mathbf{b}'||\mathbf{c}'|}\right) = \cos^{-1}\left(\frac{a^{-2}(-1 - 1 + 1)}{(\sqrt{3}\,a^{-1})^2}\right) = \cos^{-1}(-\tfrac{1}{3}) = 109.5°.$$

Other pairs yield the same value.

(c) Planes with normals in the $\mathbf{k}$ direction are clearly separated by $\frac{1}{2}a$.

A plane with its normal in the direction $\mathbf{i} + \mathbf{j}$ has an equation of the form

$$\frac{1}{\sqrt{2}}(1, 1, 0) \cdot (x, y, z) = p,$$

where p is the perpendicular distance of the origin from the plane. Since the plane with the smallest positive value of p passes through $(\frac{1}{2}a, 0, \frac{1}{2}a)$, p has the value $a/\sqrt{8}$, which is therefore the distance between successive planes with normals in the direction $\mathbf{i} + \mathbf{j}$.

Planes with their normals in the direction $\mathbf{i} + \mathbf{j} + \mathbf{k}$ have equations of the form

$$\frac{1}{\sqrt{3}}(1, 1, 1) \cdot (x, y, z) = p.$$

For the plane P_1 containing $\mathbf{b}$, $\mathbf{c}$ and $\mathbf{d}$ we have (for $\mathbf{b}$, say)

$$\frac{1}{\sqrt{3}}(1, 1, 1) \cdot (0, \tfrac{1}{2}a, \tfrac{1}{2}a) = p_1,$$

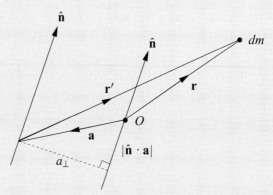

Figure 9.2 The vectors used in the proof of the parallel axis theorem in Problem 9.23.

giving $p_1 = a/\sqrt{3}$. Similarly, for the plane P_2 containing $\mathbf{c} + \mathbf{d}$, $\mathbf{b} + \mathbf{d}$ and $\mathbf{b} + \mathbf{c}$ we have (for $\mathbf{c} + \mathbf{d}$, say)

$$\frac{1}{\sqrt{3}}(1,\,1,\,1) \cdot (a,\,\tfrac{1}{2}a,\,\tfrac{1}{2}a) = p_2,$$

giving $p_2 = 2a/\sqrt{3}$. Thus the distance, d, between successive planes with normals in the direction $\mathbf{i} + \mathbf{j} + \mathbf{k}$ is the difference between these two values, i.e. $d = p_2 - p_1 = a/\sqrt{3}$.

9.23 By proceeding as indicated below, prove the *parallel axis theorem*, which states that, for a body of mass M, the moment of inertia I about any axis is related to the corresponding moment of inertia I_0 about a parallel axis that passes through the centre of mass of the body by

$$I = I_0 + Ma_{\perp}^2,$$

where $a_{\perp}$ is the perpendicular distance between the two axes. Note that I_0 can be written as

$$\int (\hat{\mathbf{n}} \times \mathbf{r}) \cdot (\hat{\mathbf{n}} \times \mathbf{r}) \, dm,$$

where $\mathbf{r}$ is the vector position, relative to the centre of mass, of the infinitesimal mass dm and $\hat{\mathbf{n}}$ is a unit vector in the direction of the axis of rotation. Write a similar expression for I in which $\mathbf{r}$ is replaced by $\mathbf{r}' = \mathbf{r} - \mathbf{a}$, where $\mathbf{a}$ is the vector position of any point on the axis to which I refers. Use Lagrange's identity and the fact that $\int \mathbf{r} \, dm = \mathbf{0}$ (by the definition of the centre of mass) to establish the result.

Figure 9.2 shows the vectors involved in describing the physical arrangement. With

$$I_0 = \int (\hat{\mathbf{n}} \times \mathbf{r}) \cdot (\hat{\mathbf{n}} \times \mathbf{r}) \, dm$$

$$= \int \left[(\hat{\mathbf{n}} \cdot \hat{\mathbf{n}})(\mathbf{r} \cdot \mathbf{r}) - (\hat{\mathbf{n}} \cdot \mathbf{r})^2 \right] dm,$$

the moment of inertia of the same mass distribution about a parallel axis passing through **a** is given by

$$I = \int (\hat{\mathbf{n}} \times \mathbf{r}') \cdot (\hat{\mathbf{n}} \times \mathbf{r}') \, dm$$

$$= \int [\hat{\mathbf{n}} \times (\mathbf{r} - \mathbf{a})] \cdot [\hat{\mathbf{n}} \times (\mathbf{r} - \mathbf{a})] \, dm$$

$$= \int \left\{ (\hat{\mathbf{n}} \cdot \hat{\mathbf{n}})[(\mathbf{r} - \mathbf{a}) \cdot (\mathbf{r} - \mathbf{a})] - [\hat{\mathbf{n}} \cdot (\mathbf{r} - \mathbf{a})]^2 \right\} \, dm,$$

$$= \int \left[r^2 - 2\mathbf{a} \cdot \mathbf{r} + a^2 - (\hat{\mathbf{n}} \cdot \mathbf{r})^2 + 2(\hat{\mathbf{n}} \cdot \mathbf{r})(\hat{\mathbf{n}} \cdot \mathbf{a}) - (\hat{\mathbf{n}} \cdot \mathbf{a})^2 \right] \, dm$$

$$= I_0 - 2\mathbf{a} \cdot \mathbf{0} + 2(\hat{\mathbf{n}} \cdot \mathbf{a})(\hat{\mathbf{n}} \cdot \mathbf{0}) + \int \left[a^2 - (\hat{\mathbf{n}} \cdot \mathbf{a})^2 \right] \, dm$$

$$= I_0 + a_\perp^2 M.$$

When obtaining the penultimate line we (twice) used the fact that O is the centre of mass of the body and so, by definition, $\int \mathbf{r} \, dm = \mathbf{0}$. To obtain the final line we noted that $\hat{\mathbf{n}} \cdot \mathbf{a}$ is the component of **a** parallel to $\hat{\mathbf{n}}$ and so $a^2 - (\hat{\mathbf{n}} \cdot \mathbf{a})^2$ is the square of the component of **a** perpendicular to $\hat{\mathbf{n}}$.

9.25 Define a set of (non-orthogonal) base vectors $\mathbf{a} = \mathbf{j} + \mathbf{k}$, $\mathbf{b} = \mathbf{i} + \mathbf{k}$ and $\mathbf{c} = \mathbf{i} + \mathbf{j}$.

(a) Establish their reciprocal vectors and hence express the vectors $\mathbf{p} = 3\mathbf{i} - 2\mathbf{j} + \mathbf{k}$, $\mathbf{q} = \mathbf{i} + 4\mathbf{j}$ and $\mathbf{r} = -2\mathbf{i} + \mathbf{j} + \mathbf{k}$ in terms of the base vectors **a**, **b** and **c**.
(b) Verify that the scalar product $\mathbf{p} \cdot \mathbf{q}$ has the same value, -5, when evaluated using either set of components.

The new base vectors are $\mathbf{a} = (0, 1, 1)$, $\mathbf{b} = (1, 0, 1)$ and $\mathbf{c} = (1, 1, 0)$.
  (a) The corresponding reciprocal vectors are thus

$$\mathbf{a}' = \frac{\mathbf{b} \times \mathbf{c}}{[\mathbf{a}, \mathbf{b}, \mathbf{c}]} = \frac{(-1, 1, 1)}{2} = \tfrac{1}{2}(-1, 1, 1),$$

and similarly for $\mathbf{b}' = \tfrac{1}{2}(1, -1, 1)$ and $\mathbf{c}' = \tfrac{1}{2}(1, 1, -1)$.
  The coefficient of (say) **a** in the expression for (say) **p** is $\mathbf{a}' \cdot \mathbf{p} = -2$. The coefficient of **b** is $\mathbf{b}' \cdot \mathbf{p} = 3$, etc. Building up each of **p**, **q** and **r** in this way, we find that their coordinates in terms of the new basis $\{\mathbf{a}, \mathbf{b}, \mathbf{c}\}$ are $\mathbf{p} = (-2, 3, 0)$, $\mathbf{q} = (\tfrac{3}{2}, -\tfrac{3}{2}, \tfrac{5}{2})$ and $\mathbf{r} = (2, -1, -1)$.
  (b) The new basis vectors, which are neither orthogonal nor normalised, have the properties $\mathbf{a} \cdot \mathbf{a} = \mathbf{b} \cdot \mathbf{b} = \mathbf{c} \cdot \mathbf{c} = 2$ and $\mathbf{b} \cdot \mathbf{c} = \mathbf{c} \cdot \mathbf{a} = \mathbf{a} \cdot \mathbf{b} = 1$. Thus the scalar product $\mathbf{p} \cdot \mathbf{q}$, calculated in the new basis, has the value

$$2 \left(-3 - \tfrac{9}{2} + 0 \right) + 1 \left(3 - 5 + \tfrac{9}{2} + \tfrac{15}{2} + 0 + 0 \right) = -15 + 10 = -5.$$

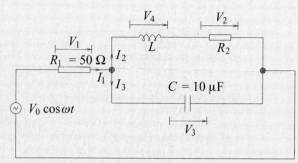

Figure 9.3 The oscillatory electric circuit in Problem 9.27. The power supply has angular frequency $\omega = 2\pi f = 400\pi \text{ s}^{-1}$.

Using the original basis, $\mathbf{p} \cdot \mathbf{q} = 3 - 8 + 0 = -5$, verifying that the scalar product has the same value in both sets of coordinates.

9.27 According to alternating current theory, the currents and potential differences in the components of the circuit shown in Figure 9.3 are determined by Kirchhoff's laws and the relationships

$$I_1 = \frac{V_1}{R_1}, \qquad I_2 = \frac{V_2}{R_2}, \qquad I_3 = i\omega C V_3, \qquad V_4 = i\omega L I_2.$$

The factor $i = \sqrt{-1}$ in the expression for I_3 indicates that the phase of I_3 is 90° ahead of V_3. Similarly, the phase of V_4 is 90° ahead of I_2.

Measurement shows that V_3 has an amplitude of $0.661 V_0$ and a phase of $+13.4°$ relative to that of the power supply. Taking $V_0 = 1 \text{ V}$ and using a series of vector plots for potential differences and currents (they could all be on the same plot if suitable scales were chosen), determine all unknown currents and potential differences and find values for the inductance of L and the resistance of R_2.

[Scales of 1 cm = 0.1 V for potential differences and 1 cm = 1 mA for currents are convenient.]

Using the suggested scales, we construct the vectors shown in Figure 9.4 in the following order:

(1) V_0 joining $(0, 0)$ to $(10, 0)$;
(2) V_3 of length 6.61 and phase $+13.4°$;
(3) $V_1 = V_0 - V_3$;
(4) I_1 parallel to V_1 and $(0.1 \times 1000)/50 = 2$ times as long;
(5) I_3, 90° ahead of V_3 in phase and $(0.1 \times 1000) \times 400\pi \times 10^{-5} = 1.26$ times as long;
(6) $I_2 = I_1 - I_3$;
(7) draw a parallel to I_2 through the origin;
(8) drop a perpendicular from V_3 onto this parallel to I_2;
(9) since $V_3 = V_2 + V_4$ and $V_2 \parallel I_2$, whilst $V_4 \perp I_2$, the foot of the perpendicular gives V_2;
(10) $V_4 = V_3 - V_2$.

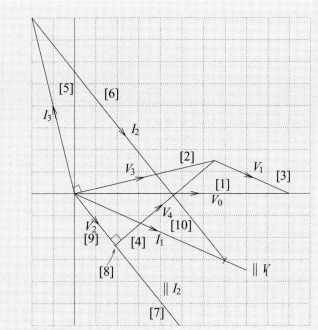

Figure 9.4 The vector solution to Problem 9.27.

The corresponding steps are labelled in the figure, which is somewhat reduced from its actual size.

Finally, $R_2 = V_2/I_2$ and $L = (V_4 \times 0.1 \times 1000)/(400\pi \times I_2)$.

The accurate solutions (obtained by calculation rather than drawing) are:

$$I_1 = (7.76, -23.2°), I_2 = (14.36, -50.8°), I_3 = (8.30, 103.4°);$$

$$V_1 = (0.388, -23.2°), V_2 = (0.287, -50.8°), V_4 = (0.596, 39.2°);$$

$$L = 33 \text{ mH}, R_2 = 20 \ \Omega.$$

10

Matrices and vector spaces

10.1 Which of the following statements about linear vector spaces are true? Where a statement is false, give a counter-example to demonstrate this.

(a) Non-singular $N \times N$ matrices form a vector space of dimension N^2.
(b) Singular $N \times N$ matrices form a vector space of dimension N^2.
(c) Complex numbers form a vector space of dimension 2.
(d) Polynomial functions of x form an infinite-dimensional vector space.
(e) Series $\{a_0, a_1, a_2, \ldots, a_N\}$ for which $\sum_{n=0}^{N} |a_n|^2 = 1$ form an N-dimensional vector space.
(f) Absolutely convergent series form an infinite-dimensional vector space.
(g) Convergent series with terms of alternating sign form an infinite-dimensional vector space.

We first remind ourselves that for a set of entities to form a vector space they must pass five tests: (i) closure under commutative and associative addition; (ii) closure under multiplication by a scalar; (iii) the existence of a null vector in the set; (iv) multiplication by unity leaves any vector unchanged; (v) each vector has a corresponding negative vector.

(a) False. The matrix $\mathbf{0}_N$, the $N \times N$ null matrix, required by (iii) is *not* non-singular and is therefore not in the set.

(b) Consider the sum of $\begin{pmatrix} 1 & 0 \\ 0 & 0 \end{pmatrix}$ and $\begin{pmatrix} 0 & 0 \\ 0 & 1 \end{pmatrix}$. The sum is the unit matrix which is not singular and so the set is not closed; this violates requirement (i). The statement is false.

(c) The space is closed under addition and multiplication by a scalar; multiplication by unity leaves a complex number unchanged; there is a null vector ($= 0 + i0$) and a negative complex number for each vector. All the necessary conditions are satisfied and the statement is true.

(d) As in the previous case, all the conditions are satisfied and the statement is true.

(e) This statement is false. To see why, consider $b_n = a_n + a_n$ for which $\sum_{n=0}^{N} |b_n|^2 = 4 \neq 1$, i.e. the set is not closed (violating (i)), or note that there is no zero vector with unit norm (violating (iii)).

(f) True. Note that an absolutely convergent series remains absolutely convergent when the signs of all of its terms are reversed.

(g) False. Consider the two series defined by

$$a_0 = \tfrac{1}{2}, \quad a_n = 2(-\tfrac{1}{2})^n \text{ for } n \geq 1; \quad b_n = -(-\tfrac{1}{2})^n \text{ for } n \geq 0.$$

The series that is the sum of $\{a_n\}$ and $\{b_n\}$ does not have alternating signs and so closure (required by (i)) does not hold.

10.3 By considering the matrices

$$A = \begin{pmatrix} 1 & 0 \\ 0 & 0 \end{pmatrix}, \qquad B = \begin{pmatrix} 0 & 0 \\ 3 & 4 \end{pmatrix},$$

show that $AB = 0$ does *not* imply that either A or B is the zero matrix but that it does imply that at least one of them is singular.

We have

$$AB = \begin{pmatrix} 1 & 0 \\ 0 & 0 \end{pmatrix} \begin{pmatrix} 0 & 0 \\ 3 & 4 \end{pmatrix} = \begin{pmatrix} 0 & 0 \\ 0 & 0 \end{pmatrix}.$$

Thus AB is the zero matrix 0 without either $A = 0$ or $B = 0$.

However, $AB = 0 \Rightarrow |A||B| = |0| = 0$ and therefore either $|A| = 0$ or $|B| = 0$ (or both).

10.5 Using the properties of determinants, solve with a minimum of calculation the following equations for x:

(a) $\begin{vmatrix} x & a & a & 1 \\ a & x & b & 1 \\ a & b & x & 1 \\ a & b & c & 1 \end{vmatrix} = 0,$ (b) $\begin{vmatrix} x+2 & x+4 & x-3 \\ x+3 & x & x+5 \\ x-2 & x-1 & x+1 \end{vmatrix} = 0.$

(a) In view of the similarities between some rows and some columns, the property most likely to be useful here is that if a determinant has two rows/columns equal (or multiples of each other) then its value is zero.

  (i) We note that setting $x = a$ makes the first and fourth columns multiples of each other and hence makes the value of the determinant 0; thus $x = a$ is one solution to the equation.

 (ii) Setting $x = b$ makes the second and third rows equal, and again the determinant vanishes; thus b is another root of the equation.

(iii) Setting $x = c$ makes the third and fourth rows equal, and yet again the determinant vanishes; thus c is also a root of the equation.

Since the determinant contains no x in its final column, it is a cubic polynomial in x and there will be exactly three roots to the equation. We have already found all three!

(b) Here, the presence of x multiplied by unity in every entry means that subtracting rows/columns will lead to a simplification. After (i) subtracting the first column from each of the others and then (ii) subtracting the first row from each of the others, the determinant becomes

$$\begin{vmatrix} x+2 & 2 & -5 \\ x+3 & -3 & 2 \\ x-2 & 1 & 3 \end{vmatrix} = \begin{vmatrix} x+2 & 2 & -5 \\ 1 & -5 & 7 \\ -4 & -1 & 8 \end{vmatrix}$$

$$= (x+2)(-40+7) + 2(-28-8) - 5(-1-20)$$

$$= -33(x+2) - 72 + 105$$

$$= -33x - 33.$$

Thus $x = -1$ is the only solution to the original (linear!) equation.

10.7 Prove the following results involving Hermitian matrices.

(a) If A is Hermitian and U is unitary then $U^{-1}AU$ is Hermitian.
(b) If A is anti-Hermitian then iA is Hermitian.
(c) The product of two Hermitian matrices A and B is Hermitian if and only if A and B commute.
(d) If S is a real antisymmetric matrix then $A = (I - S)(I + S)^{-1}$ is orthogonal. If A is given by

$$A = \begin{pmatrix} \cos\theta & \sin\theta \\ -\sin\theta & \cos\theta \end{pmatrix}$$

then find the matrix S that is needed to express A in the above form.
(e) If K is skew-Hermitian, i.e. $K^\dagger = -K$, then $V = (I + K)(I - K)^{-1}$ is unitary.

The general properties of matrices that we will need are $(A^\dagger)^{-1} = (A^{-1})^\dagger$ and

$$(AB\cdots C)^\mathrm{T} = C^\mathrm{T}\cdots B^\mathrm{T}A^\mathrm{T}, \qquad (AB\cdots C)^\dagger = C^\dagger\cdots B^\dagger A^\dagger.$$

(a) Given that $A = A^\dagger$ and $U^\dagger U = I$, consider

$$(U^{-1}AU)^\dagger = U^\dagger A^\dagger (U^{-1})^\dagger = U^{-1}A(U^\dagger)^{-1} = U^{-1}A(U^{-1})^{-1} = U^{-1}AU,$$

i.e. $U^{-1}AU$ is Hermitian.
    (b) Given $A^\dagger = -A$, consider

$$(iA)^\dagger = -iA^\dagger = -i(-A) = iA,$$

i.e. iA is Hermitian.
    (c) Given $A = A^\dagger$ and $B = B^\dagger$.

(i) Suppose $AB = BA$, then

$$(AB)^\dagger = B^\dagger A^\dagger = BA = AB,$$

i.e. AB is Hermitian.
    (ii) Now suppose that $(AB)^\dagger = AB$. Then

$$BA = B^\dagger A^\dagger = (AB)^\dagger = AB,$$

i.e. A and B commute.

Thus, AB is Hermitian $\Longleftrightarrow$ A and B commute.

(d) Given that S is real and $S^T = -S$ with $A = (I - S)(I + S)^{-1}$, consider

$$
\begin{aligned}
A^T A &= [(I - S)(I + S)^{-1}]^T [(I - S)(I + S)^{-1}] \\
&= [(I + S)^{-1}]^T (I + S)(I - S)(I + S)^{-1} \\
&= (I - S)^{-1}(I + S - S - S^2)(I + S)^{-1} \\
&= (I - S)^{-1}(I - S)(I + S)(I + S)^{-1} \\
&= II = I,
\end{aligned}
$$

i.e. A is orthogonal.

If $A = (I - S)(I + S)^{-1}$, then $A + AS = I - S$ and $(A + I)S = I - A$, giving

$$S = (A + I)^{-1}(I - A)$$

$$
= \begin{pmatrix} 1 + \cos\theta & \sin\theta \\ -\sin\theta & 1 + \cos\theta \end{pmatrix}^{-1} \begin{pmatrix} 1 - \cos\theta & -\sin\theta \\ \sin\theta & 1 - \cos\theta \end{pmatrix}
$$

$$
= \frac{1}{2 + 2\cos\theta} \begin{pmatrix} 1 + \cos\theta & -\sin\theta \\ \sin\theta & 1 + \cos\theta \end{pmatrix} \begin{pmatrix} 1 - \cos\theta & -\sin\theta \\ \sin\theta & 1 - \cos\theta \end{pmatrix}
$$

$$
= \frac{1}{4\cos^2(\theta/2)} \begin{pmatrix} 0 & -2\sin\theta \\ 2\sin\theta & 0 \end{pmatrix}
$$

$$
= \begin{pmatrix} 0 & -\tan(\theta/2) \\ \tan(\theta/2) & 0 \end{pmatrix}.
$$

(e) This proof is almost identical to the first section of part (d) but with S replaced by $-K$ and transposed matrices replaced by Hermitian conjugate matrices.

10.9 The *commutator* $[X, Y]$ of two matrices is defined by the equation

$$[X, Y] = XY - YX.$$

Two anticommuting matrices A and B satisfy

$$A^2 = I, \quad B^2 = I, \quad [A, B] = 2iC.$$

(a) Prove that $C^2 = I$ and that $[B, C] = 2iA$.
(b) Evaluate $[[[A, B], [B, C]], [A, B]]$.

(a) From $AB - BA = 2iC$ and $AB = -BA$ it follows that $AB = iC$. Thus,

$$-C^2 = iCiC = ABAB = A(-AB)B = -(AA)(BB) = -II = -I,$$

i.e. $C^2 = I$. In deriving the above result we have used the associativity of matrix multiplication.

For the commutator of B and C,

$$[\,\mathsf{B},\mathsf{C}\,] = \mathsf{BC} - \mathsf{CB}$$
$$= \mathsf{B}(-i\mathsf{AB}) - (-i)\mathsf{ABB}$$
$$= -i(\mathsf{BA})\mathsf{B} + i\mathsf{AI}$$
$$= -i(-\mathsf{AB})\mathsf{B} + i\mathsf{A}$$
$$= i\mathsf{A} + i\mathsf{A} = 2i\mathsf{A}.$$

(b) To evaluate this multiple-commutator expression we must work outwards from the innermost 'explicit' commutators. There are three such commutators at the first stage. We also need the result that $[\,\mathsf{C},\mathsf{A}\,] = 2i\mathsf{B}$; this can be proved in the same way as that for $[\,\mathsf{B},\mathsf{C}\,]$ in part (a), or by making the cyclic replacements $A \to B \to C \to A$ in the assumptions and their consequences, as proved in part (a). Then we have

$$[\,[\,[\,\mathsf{A},\mathsf{B}\,],[\,\mathsf{B},\mathsf{C}\,]\,],[\,\mathsf{A},\mathsf{B}\,]\,] = [\,[\,2i\mathsf{C},2i\mathsf{A}\,],2i\mathsf{C}\,]$$
$$= -4[\,[\,\mathsf{C},\mathsf{A}\,],2i\mathsf{C}\,]$$
$$= -4[\,2i\mathsf{B},2i\mathsf{C}\,]$$
$$= (-4)(-4)[\,\mathsf{B},\mathsf{C}\,] = 32i\mathsf{A}.$$

10.11 A general triangle has angles α, β and γ and corresponding opposite sides a, b and c. Express the length of each side in terms of the lengths of the other two sides and the relevant cosines, writing the relationships in matrix and vector form, using the vectors having components a, b, c and $\cos\alpha$, $\cos\beta$, $\cos\gamma$. Invert the matrix and hence deduce the cosine-law expressions involving α, β and γ.

By considering each side of the triangle as the sum of the projections onto it of the other two sides, we have the three simultaneous equations:

$$a = b\cos\gamma + c\cos\beta,$$
$$b = c\cos\alpha + a\cos\gamma,$$
$$c = b\cos\alpha + a\cos\beta.$$

Written in matrix and vector form, $\mathsf{Ax} = \mathsf{y}$, they become

$$\begin{pmatrix} 0 & c & b \\ c & 0 & a \\ b & a & 0 \end{pmatrix} \begin{pmatrix} \cos\alpha \\ \cos\beta \\ \cos\gamma \end{pmatrix} = \begin{pmatrix} a \\ b \\ c \end{pmatrix}.$$

The matrix A is non-singular, since $|A| = 2abc \neq 0$, and therefore has an inverse given by

$$A^{-1} = \frac{1}{2abc} \begin{pmatrix} -a^2 & ab & ac \\ ab & -b^2 & bc \\ ac & bc & -c^2 \end{pmatrix}.$$

And so, writing $x = A^{-1}y$, we have

$$\begin{pmatrix} \cos\alpha \\ \cos\beta \\ \cos\gamma \end{pmatrix} = \frac{1}{2abc} \begin{pmatrix} -a^2 & ab & ac \\ ab & -b^2 & bc \\ ac & bc & -c^2 \end{pmatrix} \begin{pmatrix} a \\ b \\ c \end{pmatrix}.$$

From this we can read off the cosine-law equation

$$\cos\alpha = \frac{1}{2abc}(-a^3 + ab^2 + ac^2) = \frac{b^2 + c^2 - a^2}{2bc},$$

and the corresponding expressions for $\cos\beta$ and $\cos\gamma$.

10.13 Determine which of the matrices below are mutually commuting, and, for those that are, demonstrate that they have a complete set of eigenvectors in common:

$$A = \begin{pmatrix} 6 & -2 \\ -2 & 9 \end{pmatrix}, \quad B = \begin{pmatrix} 1 & 8 \\ 8 & -11 \end{pmatrix},$$

$$C = \begin{pmatrix} -9 & -10 \\ -10 & 5 \end{pmatrix}, \quad D = \begin{pmatrix} 14 & 2 \\ 2 & 11 \end{pmatrix}.$$

To establish the result we need to examine all pairs of products.

$$AB = \begin{pmatrix} 6 & -2 \\ -2 & 9 \end{pmatrix} \begin{pmatrix} 1 & 8 \\ 8 & -11 \end{pmatrix}$$

$$= \begin{pmatrix} -10 & 70 \\ 70 & -115 \end{pmatrix}$$

$$= \begin{pmatrix} 1 & 8 \\ 8 & -11 \end{pmatrix} \begin{pmatrix} 6 & -2 \\ -2 & 9 \end{pmatrix} = BA.$$

$$AC = \begin{pmatrix} 6 & -2 \\ -2 & 9 \end{pmatrix} \begin{pmatrix} -9 & -10 \\ -10 & 5 \end{pmatrix}$$

$$= \begin{pmatrix} -34 & -70 \\ -72 & 65 \end{pmatrix} \neq \begin{pmatrix} -34 & -72 \\ -70 & 65 \end{pmatrix}$$

$$= \begin{pmatrix} -9 & -10 \\ -10 & 5 \end{pmatrix} \begin{pmatrix} 6 & -2 \\ -2 & 9 \end{pmatrix} = CA.$$

Continuing in this way, we find:

$$AD = \begin{pmatrix} 80 & -10 \\ -10 & 95 \end{pmatrix} = DA.$$

$$BC = \begin{pmatrix} -89 & 30 \\ 38 & -135 \end{pmatrix} \neq \begin{pmatrix} -89 & 38 \\ 30 & -135 \end{pmatrix} = CB.$$

$$BD = \begin{pmatrix} 30 & 90 \\ 90 & -105 \end{pmatrix} = DB.$$

$$CD = \begin{pmatrix} -146 & -128 \\ -130 & 35 \end{pmatrix} \neq \begin{pmatrix} -146 & -130 \\ -128 & 35 \end{pmatrix} = DC.$$

These results show that whilst A, B and D are mutually commuting, none of them commutes with C.

We could use any of the three mutually commuting matrices to find the common set (actually a pair, as they are 2×2 matrices) of eigenvectors. We arbitrarily choose A. The eigenvalues of A satisfy

$$\begin{vmatrix} 6 - \lambda & -2 \\ -2 & 9 - \lambda \end{vmatrix} = 0,$$

$$\lambda^2 - 15\lambda + 50 = 0,$$

$$(\lambda - 5)(\lambda - 10) = 0.$$

For $\lambda = 5$, an eigenvector $(x \quad y)^T$ must satisfy $x - 2y = 0$, whilst, for $\lambda = 10, 4x + 2y = 0$. Thus a pair of independent eigenvectors of A are $(2 \quad 1)^T$ and $(1 \quad -2)^T$. Direct substitution verifies that they are also eigenvectors of B and D with pairs of eigenvalues $5, -15$ and $15, 10$, respectively.

10.15 Solve the simultaneous equations

$$2x + 3y + z = 11,$$

$$x + y + z = 6,$$

$$5x - y + 10z = 34.$$

To eliminate z, (i) subtract the second equation from the first and (ii) subtract 10 times the second equation from the third.

$$x + 2y = 5,$$

$$-5x - 11y = -26.$$

To eliminate x add five times the first equation to the second

$$-y = -1.$$

Thus $y = 1$ and, by resubstitution, $x = 3$ and $z = 2$.

10.17 Show that the following equations have solutions only if $\eta = 1$ or 2, and find them in these cases:

$$x + y + z = 1, \qquad \text{(i)}$$
$$x + 2y + 4z = \eta, \qquad \text{(ii)}$$
$$x + 4y + 10z = \eta^2. \qquad \text{(iii)}$$

Expressing the equations in the form $\mathsf{A}\mathbf{x} = \mathbf{b}$, we first need to evaluate $|\mathsf{A}|$ as a preliminary to determining A^{-1}. However, we find that $|\mathsf{A}| = 1(20 - 16) + 1(4 - 10) + 1(4 - 2) = 0$. This result implies both that A is singular and has no inverse, and that the equations must be linearly dependent.

Either by observation or by solving for the combination coefficients, we see that for the LHS this linear dependence is expressed by

$$2 \times \text{(i)} + 1 \times \text{(iii)} - 3 \times \text{(ii)} = 0.$$

For a consistent solution, this must also be true for the RHSs, i.e.

$$2 + \eta^2 - 3\eta = 0.$$

This quadratic equation has solutions $\eta = 1$ and $\eta = 2$, which are therefore the only values of η for which the original equations have a solution. As the equations are linearly dependent, we may use any two to find these allowed solutions; for simplicity we use the first two in each case.

For $\eta = 1$,

$$x + y + z = 1, \quad x + 2y + 4z = 1 \Rightarrow \mathbf{x}^1 = (1 + 2\alpha \quad -3\alpha \quad \alpha)^{\mathrm{T}}.$$

For $\eta = 2$,

$$x + y + z = 1, \quad x + 2y + 4z = 2 \Rightarrow \mathbf{x}^2 = (2\alpha \quad 1 - 3\alpha \quad \alpha)^{\mathrm{T}}.$$

In both cases there is an infinity of solutions as α may take any finite value.

10.19 Make an LU decomposition of the matrix

$$\mathsf{A} = \begin{pmatrix} 3 & 6 & 9 \\ 1 & 0 & 5 \\ 2 & -2 & 16 \end{pmatrix}$$

and hence solve $\mathsf{A}\mathbf{x} = \mathbf{b}$, where (i) $\mathbf{b} = (21 \quad 9 \quad 28)^{\mathrm{T}}$, (ii) $\mathbf{b} = (21 \quad 7 \quad 22)^{\mathrm{T}}$.

Using the notation

$$\mathsf{A} = \begin{pmatrix} 1 & 0 & 0 \\ L_{21} & 1 & 0 \\ L_{31} & L_{32} & 1 \end{pmatrix} \begin{pmatrix} U_{11} & U_{12} & U_{13} \\ 0 & U_{22} & U_{23} \\ 0 & 0 & U_{33} \end{pmatrix},$$

and considering rows and columns alternately in the usual way for an LU decomposition, we require the following to be satisfied.

1st row: $U_{11} = 3$, $U_{12} = 6$, $U_{13} = 9$.
1st col: $L_{21}U_{11} = 1$, $L_{31}U_{11} = 2 \Rightarrow L_{21} = \frac{1}{3}$, $L_{31} = \frac{2}{3}$.
2nd row: $L_{21}U_{12} + U_{22} = 0$, $L_{21}U_{13} + U_{23} = 5 \Rightarrow U_{22} = -2$, $U_{23} = 2$.
2nd col: $L_{31}U_{12} + L_{32}U_{22} = -2 \Rightarrow L_{32} = 3$.
3rd row: $L_{31}U_{13} + L_{32}U_{23} + U_{33} = 16 \Rightarrow U_{33} = 4$.

Thus

$$L = \begin{pmatrix} 1 & 0 & 0 \\ \frac{1}{3} & 1 & 0 \\ \frac{2}{3} & 3 & 1 \end{pmatrix} \quad \text{and} \quad U = \begin{pmatrix} 3 & 6 & 9 \\ 0 & -2 & 2 \\ 0 & 0 & 4 \end{pmatrix}.$$

To solve $Ax = b$ with $A = LU$, we first determine y from $Ly = b$ and then solve $Ux = y$ for x.

(i) For $Ax = (21 \quad 9 \quad 28)^T$, we first solve

$$\begin{pmatrix} 1 & 0 & 0 \\ \frac{1}{3} & 1 & 0 \\ \frac{2}{3} & 3 & 1 \end{pmatrix} \begin{pmatrix} y_1 \\ y_2 \\ y_3 \end{pmatrix} = \begin{pmatrix} 21 \\ 9 \\ 28 \end{pmatrix}.$$

This can be done, almost by inspection, to give $y = (21 \quad 2 \quad 8)^T$.
We can now write $Ux = y$ explicitly as

$$\begin{pmatrix} 3 & 6 & 9 \\ 0 & -2 & 2 \\ 0 & 0 & 4 \end{pmatrix} \begin{pmatrix} x_1 \\ x_2 \\ x_3 \end{pmatrix} = \begin{pmatrix} 21 \\ 2 \\ 8 \end{pmatrix}$$

to give, equally easily, that the solution to the original matrix equation is $x = (-1 \quad 1 \quad 2)^T$.

(ii) To solve $Ax = (21 \quad 7 \quad 22)^T$ we use exactly the same forms for L and U, but the new values for the components of b, to obtain $y = (21 \quad 0 \quad 8)^T$ leading to the solution $x = (-3 \quad 2 \quad 2)^T$.

10.21 Use the Cholesky decomposition method to determine whether the following matrices are positive definite. For each, that is, determine the corresponding lower diagonal matrix L:

$$A = \begin{pmatrix} 2 & 1 & 3 \\ 1 & 3 & -1 \\ 3 & -1 & 1 \end{pmatrix}, \quad B = \begin{pmatrix} 5 & 0 & \sqrt{3} \\ 0 & 3 & 0 \\ \sqrt{3} & 0 & 3 \end{pmatrix}.$$

The matrix A is real and so we seek a real lower-diagonal matrix L such that $LL^T = A$. In order to avoid a lot of subscripts, we use lower-case letters as the non-zero elements of L:

$$\begin{pmatrix} a & 0 & 0 \\ b & c & 0 \\ d & e & f \end{pmatrix} \begin{pmatrix} a & b & d \\ 0 & c & e \\ 0 & 0 & f \end{pmatrix} = \begin{pmatrix} 2 & 1 & 3 \\ 1 & 3 & -1 \\ 3 & -1 & 1 \end{pmatrix}.$$

Firstly, from A_{11}, $a^2 = 2$. Since an overall negative sign multiplying the elements of L is irrelevant, we may choose $a = +\sqrt{2}$. Next, $ba = A_{12} = 1$, implying that $b = 1/\sqrt{2}$. Similarly, $d = 3/\sqrt{2}$.

From the second row of A we have

$$b^2 + c^2 = 3 \quad \Rightarrow \quad c = \sqrt{\tfrac{5}{2}},$$

$$bd + ce = -1 \quad \Rightarrow \quad e = \sqrt{\tfrac{2}{5}}(-1 - \tfrac{3}{2}) = -\sqrt{\tfrac{5}{2}}.$$

And, from the final row,

$$d^2 + e^2 + f^2 = 1 \Rightarrow f = (1 - \tfrac{9}{2} - \tfrac{5}{2})^{1/2} = \sqrt{-6}.$$

That f is imaginary shows that A is not a positive definite matrix.

The corresponding argument (keeping the same symbols but with different numerical values) for the matrix B is as follows.

Firstly, from $A_{11}, a^2 = 5$. Since an overall negative sign multiplying the elements of L is irrelevant, we may choose $a = +\sqrt{5}$. Next, $ba = B_{12} = 0$, implying that $b = 0$. Similarly, $d = \sqrt{3}/\sqrt{5}$.

From the second row of B we have

$$b^2 + c^2 = 3 \quad \Rightarrow \quad c = \sqrt{3},$$

$$bd + ce = 0 \quad \Rightarrow \quad e = \sqrt{\tfrac{1}{3}}(0 - 0) = 0.$$

And, from the final row,

$$d^2 + e^2 + f^2 = 3 \Rightarrow f = (3 - \tfrac{3}{5} - 0)^{1/2} = \sqrt{\tfrac{12}{5}}.$$

Thus all the elements of L have been calculated and found to be real and, in summary,

$$L = \begin{pmatrix} \sqrt{5} & 0 & 0 \\ 0 & \sqrt{3} & 0 \\ \sqrt{\tfrac{3}{5}} & 0 & \sqrt{\tfrac{12}{5}} \end{pmatrix}.$$

That $LL^T = B$ can be confirmed by substitution.

10.23 Find three real orthogonal column matrices, each of which is a simultaneous eigenvector of

$$A = \begin{pmatrix} 0 & 0 & 1 \\ 0 & 1 & 0 \\ 1 & 0 & 0 \end{pmatrix} \quad \text{and} \quad B = \begin{pmatrix} 0 & 1 & 1 \\ 1 & 0 & 1 \\ 1 & 1 & 0 \end{pmatrix}.$$

We first note that

$$AB = \begin{pmatrix} 1 & 1 & 0 \\ 1 & 0 & 1 \\ 0 & 1 & 1 \end{pmatrix} = BA.$$

The two matrices commute and so they *will* have a common set of eigenvectors.

The eigenvalues of A are given by

$$\begin{vmatrix} -\lambda & 0 & 1 \\ 0 & 1-\lambda & 0 \\ 1 & 0 & -\lambda \end{vmatrix} = (1-\lambda)(\lambda^2 - 1) = 0,$$

i.e. $\lambda = 1$, $\lambda = 1$ and $\lambda = -1$, with corresponding eigenvectors $e^1 = (1 \quad y_1 \quad 1)^T$, $e^2 = (1 \quad y_2 \quad 1)^T$ and $e^3 = (1 \quad 0 \quad -1)^T$. For these to be mutually orthogonal requires that $y_1 y_2 = -2$.

The third vector, e^3, is clearly an eigenvector of B with eigenvalue $\mu_3 = -1$. For e^1 or e^2 to be an eigenvector of B with eigenvalue μ requires

$$\begin{pmatrix} 0-\mu & 1 & 1 \\ 1 & 0-\mu & 1 \\ 1 & 1 & 0-\mu \end{pmatrix} \begin{pmatrix} 1 \\ y \\ 1 \end{pmatrix} = \begin{pmatrix} 0 \\ 0 \\ 0 \end{pmatrix};$$

i.e.
$$-\mu + y + 1 = 0,$$

and
$$1 - \mu y + 1 = 0,$$

giving
$$-\frac{2}{y} + y + 1 = 0,$$

$$\Rightarrow \quad y^2 + y - 2 = 0,$$

$$\Rightarrow \quad y = 1 \quad \text{or} \quad -2.$$

Thus, $y_1 = 1$ with $\mu_1 = 2$, whilst $y_2 = -2$ with $\mu_2 = -1$.

The common eigenvectors are thus

$$e^1 = (1 \quad 1 \quad 1)^T, \quad e^2 = (1 \quad -2 \quad 1)^T, \quad e^3 = (1 \quad 0 \quad -1)^T.$$

We note, as a check, that $\sum_i \mu_i = 2 + (-1) + (-1) = 0 = \text{Tr } B$.

10.25 Given that A is a real symmetric matrix with normalised eigenvectors e^i, obtain the coefficients α_i involved when column matrix x, which is the solution of

$$Ax - \mu x = v, \qquad (*)$$

is expanded as $x = \sum_i \alpha_i e^i$. Here μ is a given constant and v is a given column matrix.

(a) Solve $(*)$ when

$$A = \begin{pmatrix} 2 & 1 & 0 \\ 1 & 2 & 0 \\ 0 & 0 & 3 \end{pmatrix},$$

$\mu = 2$ and $v = (1 \quad 2 \quad 3)^T$.

(b) Would $(*)$ have a solution if $\mu = 1$ and (i) $v = (1 \quad 2 \quad 3)^T$, (ii) $v = (2 \quad 2 \quad 3)^T$? Where it does, find it.

Let $\mathsf{x} = \sum_i \alpha_i \mathsf{e}^i$, where $\mathsf{A}\mathsf{e}^i = \lambda_i \mathsf{e}^i$. Then

$$\mathsf{A}\mathsf{x} - \mu \mathsf{x} = \mathsf{v},$$

$$\sum_i \mathsf{A}\alpha_i \mathsf{e}^i - \sum_i \mu \alpha_i \mathsf{e}^i = \mathsf{v},$$

$$\sum_i \left(\lambda_i \alpha_i \mathsf{e}^i - \mu \alpha_i \mathsf{e}^i \right) = \mathsf{v},$$

$$\alpha_j = \frac{(\mathsf{e}^j)^\dagger \mathsf{v}}{\lambda_j - \mu}.$$

To obtain the last line we have used the mutual orthogonality of the eigenvectors. We note, in passing, that if $\mu = \lambda_j$ for any j there is no solution unless $(\mathsf{e}^j)^\dagger \mathsf{v} = 0$.

(a) To obtain the eigenvalues of the given matrix A, consider

$$0 = |\mathsf{A} - \lambda \mathsf{I}| = (3 - \lambda)(4 - 4\lambda + \lambda^2 - 1) = (3 - \lambda)(3 - \lambda)(1 - \lambda).$$

The eigenvalues, and a possible set of corresponding normalised eigenvectors, are therefore

$$\text{for} \quad \lambda = 3, \quad \mathsf{e}^1 = (0 \quad 0 \quad 1)^\mathrm{T};$$

$$\text{for} \quad \lambda = 3, \quad \mathsf{e}^2 = 2^{-1/2}(1 \quad 1 \quad 0)^\mathrm{T};$$

$$\text{for} \quad \lambda = 1, \quad \mathsf{e}^3 = 2^{-1/2}(1 \quad -1 \quad 0)^\mathrm{T}.$$

Since $\lambda = 3$ is a degenerate eigenvalue, there are infinitely many acceptable pairs of orthogonal eigenvectors corresponding to it; any pair of vectors of the form (a_i, a_i, b_i) with $2a_1 a_2 + b_1 b_2 = 0$ will suffice. The pair given is just about the simplest choice possible.

With $\mu = 2$ and $\mathsf{v} = (1 \quad 2 \quad 3)^\mathrm{T}$,

$$\alpha_1 = \frac{3}{3 - 2}, \quad \alpha_2 = \frac{3/\sqrt{2}}{3 - 2}, \quad \alpha_3 = \frac{-1/\sqrt{2}}{1 - 2}.$$

Thus the solution vector is

$$\mathsf{x} = 3 \begin{pmatrix} 0 \\ 0 \\ 1 \end{pmatrix} + \frac{3}{\sqrt{2}} \frac{1}{\sqrt{2}} \begin{pmatrix} 1 \\ 1 \\ 0 \end{pmatrix} + \frac{1}{\sqrt{2}} \frac{1}{\sqrt{2}} \begin{pmatrix} 1 \\ -1 \\ 0 \end{pmatrix} = \begin{pmatrix} 2 \\ 1 \\ 3 \end{pmatrix}.$$

(b) If $\mu = 1$ then it is equal to the third eigenvalue and a solution is only possible if $(\mathsf{e}^3)^\dagger \mathsf{v} = 0$.

For (i) $\mathsf{v} = (1 \quad 2 \quad 3)^\mathrm{T}$, $(\mathsf{e}^3)^\dagger \mathsf{v} = -1/\sqrt{2}$ and so no solution is possible.

For (ii) $\mathsf{v} = (2 \quad 2 \quad 3)^\mathrm{T}$, $(\mathsf{e}^3)^\dagger \mathsf{v} = 0$, and so a solution is possible. The other scalar products needed are $(\mathsf{e}^1)^\dagger \mathsf{v} = 3$ and $(\mathsf{e}^2)^\dagger \mathsf{v} = 2\sqrt{2}$. For this vector v the solution to the equation is

$$\mathsf{x} = \frac{3}{3 - 1} \begin{pmatrix} 0 \\ 0 \\ 1 \end{pmatrix} + \frac{2\sqrt{2}}{3 - 1} \frac{1}{\sqrt{2}} \begin{pmatrix} 1 \\ 1 \\ 0 \end{pmatrix} = \begin{pmatrix} 1 \\ 1 \\ \frac{3}{2} \end{pmatrix}.$$

[The solutions to both parts can be checked by resubstitution.]

10.27 By finding the eigenvectors of the Hermitian matrix

$$H = \begin{pmatrix} 10 & 3i \\ -3i & 2 \end{pmatrix},$$

construct a unitary matrix U such that $U^\dagger HU = \Lambda$, where Λ is a real diagonal matrix.

We start by finding the eigenvalues of H using

$$\begin{vmatrix} 10 - \lambda & 3i \\ -3i & 2 - \lambda \end{vmatrix} = 0,$$

$$20 - 12\lambda + \lambda^2 - 3 = 0,$$

$$\lambda = 1 \quad \text{or} \quad 11.$$

As expected for an Hermitian matrix, the eigenvalues are real.

For $\lambda = 1$ and normalised eigenvector $(x \quad y)^T$,

$$9x + 3iy = 0 \quad \Rightarrow \quad x^1 = (10)^{-1/2} (1 \quad 3i)^T.$$

For $\lambda = 11$ and normalised eigenvector $(x \quad y)^T$,

$$-x + 3iy = 0 \quad \Rightarrow \quad x^2 = (10)^{-1/2} (3i \quad 1)^T.$$

Again as expected, $(x^1)^\dagger x^2 = 0$, thus verifying the mutual orthogonality of the eigenvectors. It should be noted that the normalisation factor is determined by $(x^i)^\dagger x^i = 1$ (*not* by $(x^i)^T x^i = 1$).

We now use these normalised eigenvectors of H as the columns of the matrix U and check that it is unitary:

$$U = \frac{1}{\sqrt{10}} \begin{pmatrix} 1 & 3i \\ 3i & 1 \end{pmatrix}, U^\dagger = \frac{1}{\sqrt{10}} \begin{pmatrix} 1 & -3i \\ -3i & 1 \end{pmatrix},$$

$$UU^\dagger = \frac{1}{10} \begin{pmatrix} 1 & 3i \\ 3i & 1 \end{pmatrix} \begin{pmatrix} 1 & -3i \\ -3i & 1 \end{pmatrix} = \frac{1}{10} \begin{pmatrix} 10 & 0 \\ 0 & 10 \end{pmatrix} = I.$$

U has the further property that

$$U^\dagger HU = \frac{1}{\sqrt{10}} \begin{pmatrix} 1 & -3i \\ -3i & 1 \end{pmatrix} \begin{pmatrix} 10 & 3i \\ -3i & 2 \end{pmatrix} \frac{1}{\sqrt{10}} \begin{pmatrix} 1 & 3i \\ 3i & 1 \end{pmatrix}$$

$$= \frac{1}{10} \begin{pmatrix} 1 & -3i \\ -3i & 1 \end{pmatrix} \begin{pmatrix} 1 & 33i \\ 3i & 11 \end{pmatrix}$$

$$= \frac{1}{10} \begin{pmatrix} 10 & 0 \\ 0 & 110 \end{pmatrix} = \begin{pmatrix} 1 & 0 \\ 0 & 11 \end{pmatrix} = \Lambda.$$

That the diagonal entries of Λ are the eigenvalues of H is in accord with the general theory of normal matrices.

10.29 Given that the matrix

$$A = \begin{pmatrix} 2 & -1 & 0 \\ -1 & 2 & -1 \\ 0 & -1 & 2 \end{pmatrix}$$

has two eigenvectors of the form $(1 \quad y \quad 1)^T$, use the stationary property of the expression $J(x) = x^T A x / (x^T x)$ to obtain the corresponding eigenvalues. Deduce the third eigenvalue.

Since A is real and symmetric, each eigenvalue λ is real. Further, from the first component of $Ax = \lambda x$, we have that $2 - y = \lambda$, showing that y is also real. Considered as a function of a general vector of the form $(1 \quad y \quad 1)^T$, the quadratic form $x^T A x$ can be written explicitly as

$$x^T A x = (1 \quad y \quad 1) \begin{pmatrix} 2 & -1 & 0 \\ -1 & 2 & -1 \\ 0 & -1 & 2 \end{pmatrix} \begin{pmatrix} 1 \\ y \\ 1 \end{pmatrix}$$

$$= (1 \quad y \quad 1) \begin{pmatrix} 2 - y \\ 2y - 2 \\ 2 - y \end{pmatrix}$$

$$= 2y^2 - 4y + 4.$$

The scalar product $x^T x$ has the value $2 + y^2$, and so we need to find the stationary values of

$$I = \frac{2y^2 - 4y + 4}{2 + y^2}.$$

These are given by

$$0 = \frac{dI}{dy} = \frac{(2 + y^2)(4y - 4) - (2y^2 - 4y + 4)2y}{(2 + y^2)^2}$$

$$0 = 4y^2 - 8,$$

$$y = \pm\sqrt{2}.$$

The corresponding eigenvalues are the values of I at the stationary points:

$$\text{for} \quad y = \sqrt{2}, \qquad \lambda_1 = \frac{2(2) - 4\sqrt{2} + 4}{2 + 2} = 2 - \sqrt{2};$$

$$\text{for} \quad y = -\sqrt{2}, \qquad \lambda_2 = \frac{2(2) + 4\sqrt{2} + 4}{2 + 2} = 2 + \sqrt{2}.$$

The final eigenvalue can be found using the fact that the sum of the eigenvalues is equal to the trace of the matrix; so

$$\lambda_3 = (2 + 2 + 2) - (2 - \sqrt{2}) - (2 + \sqrt{2}) = 2.$$

10.31 The equation of a particular conic section is

$$Q \equiv 8x_1^2 + 8x_2^2 - 6x_1x_2 = 110.$$

Determine the type of conic section this represents, the orientation of its principal axes and relevant lengths in the directions of these axes.

The eigenvalues of the matrix $\begin{pmatrix} 8 & -3 \\ -3 & 8 \end{pmatrix}$ associated with the quadratic form on the LHS (without any prior scaling) are given by

$$0 = \begin{vmatrix} 8 - \lambda & -3 \\ -3 & 8 - \lambda \end{vmatrix}$$

$$= \lambda^2 - 16\lambda + 55$$

$$= (\lambda - 5)(\lambda - 11).$$

Referred to the corresponding eigenvectors as axes, the conic section (an ellipse, since both eigenvalues are positive) will take the form

$$5y_1^2 + 11y_2^2 = 110 \quad \text{or, in standard form,} \quad \frac{y_1^2}{22} + \frac{y_2^2}{10} = 1.$$

Thus the semi-axes are of lengths $\sqrt{22}$ and $\sqrt{10}$; the former is in the direction of the vector $(x_1 \;\; x_2)^T$ given by $(8 - 5)x_1 - 3x_2 = 0$, i.e. it is the line $x_1 = x_2$. The other principal axis will be the line at right angles to this, namely the line $x_1 = -x_2$.

10.33 Find the direction of the axis of symmetry of the quadratic surface

$$7x^2 + 7y^2 + 7z^2 - 20yz - 20xz + 20xy = 3.$$

The straightforward, but longer, solution to this exercise is as follows.

Consider the characteristic polynomial of the matrix associated with the quadratic surface, namely

$$f(\lambda) = \begin{vmatrix} 7 - \lambda & 10 & -10 \\ 10 & 7 - \lambda & -10 \\ -10 & -10 & 7 - \lambda \end{vmatrix}$$

$$= (7 - \lambda)(-51 - 14\lambda + \lambda^2) + 10(30 + 10\lambda) - 10(-30 - 10\lambda)$$

$$= -\lambda^3 + 21\lambda^2 + 153\lambda + 243.$$

If the quadratic surface has an axis of symmetry, it must have two equal major axes (perpendicular to it), and hence the characteristic equation must have a repeated root. This same root will therefore also be a root of $df/d\lambda = 0$, i.e. of

$$-3\lambda^2 + 42\lambda + 153 = 0,$$

$$\lambda^2 - 14\lambda - 51 = 0,$$

$$\lambda = 17 \quad \text{or} \quad -3.$$

Substitution shows that -3 is a root (and therefore a double root) of $f(\lambda) = 0$, but that 17 is not. The non-repeated root can be calculated as the trace of the matrix minus the repeated roots, i.e. $21 - (-3) - (-3) = 27$. It is the eigenvector that corresponds to this eigenvalue that gives the direction $(x \quad y \quad z)^{\text{T}}$ of the axis of symmetry. Its components must satisfy

$$(7 - 27)x + 10y - 10z = 0,$$

$$10x + (7 - 27)y - 10z = 0.$$

The axis of symmetry is therefore in the direction $(1 \quad 1 \quad -1)^{\text{T}}$.

A more subtle solution is obtained by noting that setting $\lambda = -3$ makes *all three* of the rows (or columns) of the determinant multiples of each other, i.e. it reduces the determinant to rank one. Thus -3 is a repeated root of the characteristic equation and the third root is $21 - 2(-3) = 27$. The rest of the analysis is as above.

We note in passing that, as two eigenvalues are negative and equal, the surface is the hyperboloid of revolution obtained by rotating a (two-branched) hyperbola about its axis of symmetry. Referred to this axis and two others forming a mutually orthogonal set, the equation of the quadratic surface takes the form $-3\chi^2 - 3\eta^2 + 27\zeta^2 = 3$ and so the tips of the two 'nose cones' ($\chi = \eta = 0$) are separated by $\frac{2}{3}$ of a unit.

10.35 This problem demonstrates the reverse of the usual procedure of diagonalising a matrix.

(a) Rearrange the result $A' = S^{-1}AS$ (which shows how to make a change of basis that diagonalises A) so as to express the original matrix A in terms of the unitary matrix S and the diagonal matrix A'. Hence show how to construct a matrix A that has given eigenvalues and given (orthogonal) column matrices as its eigenvectors.

(b) Find the matrix that has as eigenvectors $(1 \quad 2 \quad 1)^{\text{T}}$, $(1 \quad -1 \quad 1)^{\text{T}}$ and $(1 \quad 0 \quad -1)^{\text{T}}$ and corresponding eigenvalues λ, μ and ν.

(c) Try a particular case, say $\lambda = 3$, $\mu = -2$ and $\nu = 1$, and verify by explicit solution that the matrix so found does have these eigenvalues.

(a) Since S is unitary, we can multiply the given result on the left by S and on the right by $S^{\dagger}$ to obtain

$$SA'S^{\dagger} = SS^{-1}ASS^{\dagger} = (I)A(I) = A.$$

More explicitly, in terms of the eigenvalues and normalised eigenvectors x^i of A,

$$A = (x^1 \quad x^2 \quad \cdots \quad x^n)\Lambda(x^1 \quad x^2 \quad \cdots \quad x^n)^{\dagger}.$$

Here Λ is the diagonal matrix that has the eigenvalues of A as its diagonal elements.

Now, given normalised orthogonal column matrices and n specified values, we can use this result to construct a matrix that has the column matrices as eigenvectors and the values as eigenvalues.

(b) The normalised versions of the given column vectors are

$$\frac{1}{\sqrt{6}}(1 \quad 2 \quad 1)^{\mathrm{T}}, \frac{1}{\sqrt{3}}(1 \quad -1 \quad 1)^{\mathrm{T}}, \frac{1}{\sqrt{2}}(1 \quad 0 \quad -1)^{\mathrm{T}},$$

and the orthogonal matrix S can be constructed using these as its columns:

$$\mathsf{S} = \frac{1}{\sqrt{6}} \begin{pmatrix} 1 & \sqrt{2} & \sqrt{3} \\ 2 & -\sqrt{2} & 0 \\ 1 & \sqrt{2} & -\sqrt{3} \end{pmatrix}.$$

The required matrix A can now be formed as $\mathsf{S}\Lambda\mathsf{S}^{\dagger}$:

$$\mathsf{A} = \frac{1}{6} \begin{pmatrix} 1 & \sqrt{2} & \sqrt{3} \\ 2 & -\sqrt{2} & 0 \\ 1 & \sqrt{2} & -\sqrt{3} \end{pmatrix} \begin{pmatrix} \lambda & 0 & 0 \\ 0 & \mu & 0 \\ 0 & 0 & \nu \end{pmatrix} \begin{pmatrix} 1 & 2 & 1 \\ \sqrt{2} & -\sqrt{2} & \sqrt{2} \\ \sqrt{3} & 0 & -\sqrt{3} \end{pmatrix}$$

$$= \frac{1}{6} \begin{pmatrix} 1 & \sqrt{2} & \sqrt{3} \\ 2 & -\sqrt{2} & 0 \\ 1 & \sqrt{2} & -\sqrt{3} \end{pmatrix} \begin{pmatrix} \lambda & 2\lambda & \lambda \\ \sqrt{2}\mu & -\sqrt{2}\mu & \sqrt{2}\mu \\ \sqrt{3}\nu & 0 & -\sqrt{3}\nu \end{pmatrix}$$

$$= \frac{1}{6} \begin{pmatrix} \lambda + 2\mu + 3\nu & 2\lambda - 2\mu & \lambda + 2\mu - 3\nu \\ 2\lambda - 2\mu & 4\lambda + 2\mu & 2\lambda - 2\mu \\ \lambda + 2\mu - 3\nu & 2\lambda - 2\mu & \lambda + 2\mu + 3\nu \end{pmatrix}.$$

(c) Setting $\lambda = 3$, $\mu = -2$ and $\nu = 1$, as a particular case, gives A as

$$\mathsf{A} = \frac{1}{6} \begin{pmatrix} 2 & 10 & -4 \\ 10 & 8 & 10 \\ -4 & 10 & 2 \end{pmatrix}.$$

We complete the exercise by solving for the eigenvalues of A in the usual way. To avoid working with fractions, and any confusion with the value $\lambda = 3$ used when constructing A, we will find the eigenvalues of $6\mathsf{A}$ and denote them by η.

$$0 = |6\mathsf{A} - \eta\mathsf{I}|$$

$$= \begin{vmatrix} 2 - \eta & 10 & -4 \\ 10 & 8 - \eta & 10 \\ -4 & 10 & 2 - \eta \end{vmatrix}$$

$$= (2 - \eta)(\eta^2 - 10\eta - 84) + 10(10\eta - 60) - 4(132 - 4\eta)$$

$$= -\eta^3 + 12\eta^2 + 180\eta - 1296$$

$$= -(\eta - 6)(\eta^2 - 6\eta - 216)$$

$$= -(\eta - 6)(\eta + 12)(\eta - 18).$$

Thus $6\mathsf{A}$ has eigenvalues 6, -12 and 18; the values for A itself are 1, -2 and 3, as expected.

10.37 A more general form of expression for the determinant of a 3×3 matrix A than (10.45) is given by

$$|\mathsf{A}|\epsilon_{lmn} = A_{li} A_{mj} A_{nk} \epsilon_{ijk}. \tag{10.1}$$

As can be seen by multiplying out all the brackets, (10.45) could have been written as

$$|\mathsf{A}| = \epsilon_{ijk} A_{i1} A_{j2} A_{k3}.$$

The more general form removes the explicit mention of $1, 2, 3$ at the expense of an additional Levi–Civita symbol; the form of (10.1) can be readily extended to cover a general $N \times N$ matrix.

Use this more general form to prove properties (i), (iii), (v), (vi) and (vii) of determinants stated in Section 10.7.1. Property (iv) is obvious by inspection. For definiteness take $N = 3$, but convince yourself that your methods of proof would be valid for any positive integer N.

A full account of the answer to this problem is given in the *Hints and answers* section at the end of the chapter, almost as if it were part of the main text. The reader is referred there for the details.

11

Vector calculus

11.1 A particle's position at time $t \geq 0$ is given by $\mathbf{r} = \mathbf{r}(t)$ and remains within a bounded region.

(a) By considering the time derivative of $|\mathbf{r}|^2$, show mathematically the intuitive result that when the particle is furthest from the origin its position and velocity vectors are orthogonal.

(b) For $\mathbf{r}(t) = (2\,\mathbf{i} + 5t\,\mathbf{j})e^{-t}$, find the particle's maximum distance d from the origin and verify explicitly that result (a) is valid.

(c) At what time does the particle trajectory cross the line $\mathbf{r} = \lambda(\mathbf{i} + \mathbf{j})$, and how far from the origin is it when it does so? How does the trajectory approach its end-point?

(a) The value of $|\mathbf{r}|^2$ is maximal when the derivative

$$\frac{d}{dt}|\mathbf{r}|^2 = \frac{d}{dt}(\mathbf{r} \cdot \mathbf{r}) = 2\mathbf{r} \cdot \frac{d\mathbf{r}}{dt}$$

is equal to zero, i.e. $\mathbf{r} \cdot \mathbf{v} = \mathbf{r} \cdot \dfrac{d\mathbf{r}}{dt} = 0$, showing that the position and velocity vectors are orthogonal.

(b) With $\mathbf{r}(t) = (2\,\mathbf{i} + 5t\,\mathbf{j})e^{-t}$, we have $\mathbf{r} \cdot \mathbf{r} = (4 + 25t^2)e^{-2t}$. This is maximal when

$$0 = \frac{d}{dt}(\mathbf{r} \cdot \mathbf{r}) = -2(4 + 25t^2)e^{-2t} + 50te^{-2t}.$$

Since $e^{-2t} \neq 0$, except as $t \to \infty$, we require

$$50t^2 - 50t + 8 = 0 \quad \Rightarrow \quad t + \frac{25 \pm \sqrt{625 - 400}}{50} = \frac{4}{5} \text{ or } \frac{1}{5}.$$

The corresponding distances are

$$\left(4 + \frac{25}{25}\right)^{1/2} e^{-1/5} = 1.8307 \quad \text{and} \quad \left(4 + \frac{(25)(16)}{25}\right)^{1/2} e^{-4/5} = 2.0095.$$

Clearly the maximum distance reached, at $t = 4/5$, is 2.0095.

The position vector is then $\mathbf{r} = (2\,\mathbf{i} + 4\,\mathbf{j})e^{-4/5}$. At the same time the velocity vector is

$$\mathbf{v} = \dot{\mathbf{r}} = -2e^{-t}\mathbf{i} + (5 - 5t)e^{-t}\mathbf{j} = (-2\,\mathbf{i} + \mathbf{j})e^{-4/5}.$$

Forming the scalar product of the two gives

$$\mathbf{r} \cdot \mathbf{v} = (2\,\mathbf{i} + 4\,\mathbf{j}) \cdot (-2\,\mathbf{i} + \mathbf{j})e^{-8/5} = 0,$$

showing that $\mathbf{r}$ and $\mathbf{v}$ are orthogonal.

(c) The trajectory crosses the line $\mathbf{r} = \lambda(\mathbf{i} + \mathbf{j})$ when its position coordinates, $2e^{-t}$ and $5te^{-t}$, are equal. This happens when $t = 2/5$ and the particle is then $(4 + 4)^{1/2}e^{-2/5} = 1.896$ from the origin.

As $t \to \infty$, although both coordinates $\to 0$, the $\mathbf{j}$-coordinate $(5te^{-t})$ becomes $\gg$ than the $\mathbf{i}$-coordinate $(2e^{-t})$. Thus, the trajectory approaches the origin from the direction of the positive y-axis.

11.3 Evaluate the integral

$$\int \left[\mathbf{a}(\dot{\mathbf{b}} \cdot \mathbf{a} + \mathbf{b} \cdot \dot{\mathbf{a}}) + \dot{\mathbf{a}}(\mathbf{b} \cdot \mathbf{a}) - 2(\dot{\mathbf{a}} \cdot \mathbf{a})\mathbf{b} - \dot{\mathbf{b}}|\mathbf{a}|^2 \right] dt$$

in which $\dot{\mathbf{a}}$ and $\dot{\mathbf{b}}$ are the derivatives of $\mathbf{a}$ and $\mathbf{b}$ with respect to t.

In order to evaluate this integral, we need to group the terms in the integrand so that each is a part of the total derivative of a product of factors. Clearly, the first three terms are the derivative of $\mathbf{a}(\mathbf{b} \cdot \mathbf{a})$, i.e.

$$\frac{d}{dt}[\mathbf{a}(\mathbf{b} \cdot \mathbf{a})] = \dot{\mathbf{a}}(\mathbf{b} \cdot \mathbf{a}) + \mathbf{a}(\dot{\mathbf{b}} \cdot \mathbf{a}) + \mathbf{a}(\mathbf{b} \cdot \dot{\mathbf{a}}).$$

Similarly,

$$\frac{d}{dt}[\mathbf{b}(\mathbf{a} \cdot \mathbf{a})] = \dot{\mathbf{b}}(\mathbf{a} \cdot \mathbf{a}) + \mathbf{b}(\dot{\mathbf{a}} \cdot \mathbf{a}) + \mathbf{b}(\mathbf{a} \cdot \dot{\mathbf{a}}).$$

Hence,

$$I = \int \left\{ \frac{d}{dt}[\mathbf{a}(\mathbf{b} \cdot \mathbf{a})] - \frac{d}{dt}[\mathbf{b}(\mathbf{a} \cdot \mathbf{a})] \right\} dt$$

$$= \mathbf{a}(\mathbf{b} \cdot \mathbf{a}) - \mathbf{b}(\mathbf{a} \cdot \mathbf{a}) + \mathbf{h}$$

$$= \mathbf{a} \times (\mathbf{a} \times \mathbf{b}) + \mathbf{h},$$

where $\mathbf{h}$ is the (vector) constant of integration. To obtain the final line above, we used a special case of the expansion of a vector triple product.

11.5 The general equation of motion of a (non-relativistic) particle of mass m and charge q when it is placed in a region where there is a magnetic field $\mathbf{B}$ and an electric field $\mathbf{E}$ is

$$m\ddot{\mathbf{r}} = q(\mathbf{E} + \dot{\mathbf{r}} \times \mathbf{B});$$

here $\mathbf{r}$ is the position of the particle at time t and $\dot{\mathbf{r}} = d\mathbf{r}/dt$, etc. Write this as three separate equations in terms of the Cartesian components of the vectors involved.

For the simple case of crossed uniform fields $\mathbf{E} = E\mathbf{i}$, $\mathbf{B} = B\mathbf{j}$, in which the particle starts from the origin at $t = 0$ with $\dot{\mathbf{r}} = v_0\mathbf{k}$, find the equations of motion and show the following:

(a) if $v_0 = E/B$ then the particle continues its initial motion;
(b) if $v_0 = 0$ then the particle follows the space curve given in terms of the parameter ξ by

$$x = \frac{mE}{B^2q}(1 - \cos \xi), \qquad y = 0, \qquad z = \frac{mE}{B^2q}(\xi - \sin \xi).$$

Interpret this curve geometrically and relate ξ to t. Show that the total distance travelled by the particle after time t is given by

$$\frac{2E}{B} \int_0^t \left| \sin \frac{Bqt'}{2m} \right| dt'.$$

Expressed in Cartesian coordinates, the components of the vector equation read

$$m\ddot{x} = qE_x + q(\dot{y}B_z - \dot{z}B_y),$$
$$m\ddot{y} = qE_y + q(\dot{z}B_x - \dot{x}B_z),$$
$$m\ddot{z} = qE_z + q(\dot{x}B_y - \dot{y}B_x).$$

For $E_x = E$, $B_y = B$ and all other field components zero, the equations reduce to

$$m\ddot{x} = qE - qB\dot{z}, \qquad m\ddot{y} = 0, \qquad m\ddot{z} = qB\dot{x}.$$

The second of these, together with the initial conditions $y(0) = \dot{y}(0) = 0$, implies that $y(t) = 0$ for all t. The final equation can be integrated directly to give

$$m\dot{z} = qBx + mv_0, \qquad (*)$$

which can now be substituted into the first to give a differential equation for x:

$$m\ddot{x} = qE - qB\left(\frac{qB}{m}x + v_0\right),$$

$$\Rightarrow \quad \ddot{x} + \left(\frac{qB}{m}\right)^2 x = \frac{q}{m}(E - v_0 B).$$

(a) If $v_0 = E/B$ then the equation for x is that of simple harmonic motion and

$$x(t) = A\cos\omega t + B\sin\omega t,$$

where $\omega = qB/m$. However, in the present case, the initial conditions $x(0) = \dot{x}(0) = 0$ imply that $x(t) = 0$ for all t. Thus, there is no motion in either the x- or the y-direction and, as is then shown by $(*)$, the particle continues with its initial speed v_0 in the z-direction.

(b) If $v_0 = 0$, the equation of motion is

$$\ddot{x} + \omega^2 x = \frac{qE}{m},$$

which again has sinusoidal solutions but has a non-zero RHS. The full solution consists of the same complementary function as in part (a) together with the simplest possible particular integral, namely $x = qE/m\omega^2$. It is therefore

$$x(t) = A\cos\omega t + B\sin\omega t + \frac{qE}{m\omega^2}.$$

The initial condition $x(0) = 0$ implies that $A = -qE/(m\omega^2)$, whilst $\dot{x}(0) = 0$ requires that $B = 0$. Thus,

$$x = \frac{qE}{m\omega^2}(1 - \cos\omega t),$$

$$\Rightarrow \quad \dot{z} = \frac{qB}{m}x = \omega\frac{qE}{m\omega^2}(1 - \cos\omega t) = \frac{qE}{m\omega}(1 - \cos\omega t).$$

Since $z(0) = 0$, straightforward integration gives

$$z = \frac{qE}{m\omega}\left(t - \frac{\sin\omega t}{\omega}\right) = \frac{qE}{m\omega^2}(\omega t - \sin\omega t).$$

Thus, since $qE/m\omega^2 = mE/B^2q$, the path is of the given parametric form with $\xi = \omega t$. It is a cycloid in the plane $y = 0$; the x-coordinate varies in the restricted range $0 \leq x \leq 2qE/(m\omega^2)$, whilst the z-coordinate continually increases, though not at a uniform rate.

The element of path length is given by $ds^2 = dx^2 + dy^2 + dz^2$. In this case, writing $qE/(m\omega) = E/B$ as μ,

$$ds = \left[\left(\frac{dx}{dt}\right)^2 + \left(\frac{dz}{dt}\right)^2 \right]^{1/2} dt$$

$$= \left[\mu^2 \sin^2 \omega t + \mu^2 (1 - \cos \omega t)^2 \right]^{1/2} dt$$

$$= \left[2\mu^2 (1 - \cos \omega t) \right]^{1/2} dt = 2\mu |\sin \tfrac{1}{2}\omega t| \, dt.$$

Thus the total distance travelled after time t is given by

$$s = \int_0^t 2\mu |\sin \tfrac{1}{2}\omega t'| \, dt' = \frac{2E}{B} \int_0^t \left| \sin \frac{qBt'}{2m} \right| dt'.$$

11.7 If two systems of coordinates with a common origin O are rotating with respect to each other, the measured accelerations differ in the two systems. Denoting by $\mathbf{r}$ and $\mathbf{r}'$ position vectors in frames $OXYZ$ and $OX'Y'Z'$, respectively, the connection between the two is

$$\ddot{\mathbf{r}}' = \ddot{\mathbf{r}} + \dot{\boldsymbol{\omega}} \times \mathbf{r} + 2\boldsymbol{\omega} \times \dot{\mathbf{r}} + \boldsymbol{\omega} \times (\boldsymbol{\omega} \times \mathbf{r}),$$

where $\boldsymbol{\omega}$ is the angular velocity vector of the rotation of $OXYZ$ with respect to $OX'Y'Z'$ (taken as fixed). The third term on the RHS is known as the Coriolis acceleration, whilst the final term gives rise to a centrifugal force.

Consider the application of this result to the firing of a shell of mass m from a stationary ship on the steadily rotating earth, working to the first order in ω ($= 7.3 \times 10^{-5}$ rad s^{-1}). If the shell is fired with velocity $\mathbf{v}$ at time $t = 0$ and only reaches a height that is small compared with the radius of the earth, show that its acceleration, as recorded on the ship, is given approximately by

$$\ddot{\mathbf{r}} = \mathbf{g} - 2\boldsymbol{\omega} \times (\mathbf{v} + \mathbf{g}t),$$

where $m\mathbf{g}$ is the weight of the shell measured on the ship's deck.

The shell is fired at another stationary ship (a distance $\mathbf{s}$ away) and $\mathbf{v}$ is such that the shell would have hit its target had there been no Coriolis effect.

(a) Show that without the Coriolis effect the time of flight of the shell would have been $\tau = -2\mathbf{g} \cdot \mathbf{v}/g^2$.
(b) Show further that when the shell actually hits the sea it is off-target by approximately

$$\frac{2\tau}{g^2}[(\mathbf{g} \times \boldsymbol{\omega}) \cdot \mathbf{v}](\mathbf{g}\tau + \mathbf{v}) - (\boldsymbol{\omega} \times \mathbf{v})\tau^2 - \frac{1}{3}(\boldsymbol{\omega} \times \mathbf{g})\tau^3.$$

(c) Estimate the order of magnitude Δ of this miss for a shell for which the initial speed v is 300 m s^{-1}, firing close to its maximum range ($\mathbf{v}$ makes an angle of $\pi/4$ with the vertical) in a northerly direction, whilst the ship is stationed at latitude 45° North.

As the Earth is rotating steadily $\dot{\boldsymbol{\omega}} = \mathbf{0}$, and for the mass at rest on the deck,

$$m\ddot{\mathbf{r}}' = m\mathbf{g} + \mathbf{0} + 2\boldsymbol{\omega} \times \dot{\mathbf{0}} + m\boldsymbol{\omega} \times (\boldsymbol{\omega} \times \mathbf{r}).$$

This, including the centrifugal effect, defines $\mathbf{g}$ which is assumed constant throughout the trajectory.

For the moving mass ($\ddot{\mathbf{r}}'$ is unchanged),

$$mg + \boldsymbol{\omega} \times (\boldsymbol{\omega} \times \mathbf{r}) = m\ddot{\mathbf{r}} + 2m\boldsymbol{\omega} \times \dot{\mathbf{r}} + m\boldsymbol{\omega} \times (\boldsymbol{\omega} \times \mathbf{r}),$$

i.e.

$$\ddot{\mathbf{r}} = \mathbf{g} - 2\boldsymbol{\omega} \times \dot{\mathbf{r}}.$$

Now, $\omega \dot{r} \ll g$ and so to zeroth order in ω

$$\ddot{\mathbf{r}} = \mathbf{g} \quad \Rightarrow \quad \dot{\mathbf{r}} = \mathbf{g}t + \mathbf{v}.$$

Resubstituting this into the Coriolis term gives, to first order in ω,

$$\ddot{\mathbf{r}} = \mathbf{g} - 2\boldsymbol{\omega} \times (\mathbf{v} + \mathbf{g}t).$$

(a) With no Coriolis force,

$$\dot{\mathbf{r}} = \mathbf{g}t + \mathbf{v} \quad \text{and} \quad \mathbf{r} = \tfrac{1}{2}\mathbf{g}t^2 + \mathbf{v}t.$$

Let $\mathbf{s} = \tfrac{1}{2}\mathbf{g}\tau^2 + \mathbf{v}\tau$ and use the observation that $\mathbf{s} \cdot \mathbf{g} = 0$, giving

$$\tfrac{1}{2}g^2\tau^2 + \mathbf{v} \cdot \mathbf{g}\tau = 0 \quad \Rightarrow \quad \tau = -\frac{2\mathbf{v} \cdot \mathbf{g}}{g^2}.$$

(b) With Coriolis force,

$$\ddot{\mathbf{r}} = \mathbf{g} - 2(\boldsymbol{\omega} \times \mathbf{g})t - 2(\boldsymbol{\omega} \times \mathbf{v}),$$

$$\dot{\mathbf{r}} = \mathbf{g}t - (\boldsymbol{\omega} \times \mathbf{g})t^2 - 2(\boldsymbol{\omega} \times \mathbf{v})t + \mathbf{v},$$

$$\mathbf{r} = \tfrac{1}{2}\mathbf{g}t^2 - \tfrac{1}{3}(\boldsymbol{\omega} \times \mathbf{g})t^3 - (\boldsymbol{\omega} \times \mathbf{v})t^2 + \mathbf{v}t. \qquad (*)$$

If the shell hits the sea at time T in the position $\mathbf{r} = \mathbf{s} + \boldsymbol{\Delta}$, then $(\mathbf{s} + \boldsymbol{\Delta}) \cdot \mathbf{g} = 0$, i.e.

$$0 = (\mathbf{s} + \boldsymbol{\Delta}) \cdot \mathbf{g} = \tfrac{1}{2}g^2 T^2 - 0 - (\boldsymbol{\omega} \times \mathbf{v}) \cdot \mathbf{g}\, T^2 + \mathbf{v} \cdot \mathbf{g}\, T,$$

$$\Rightarrow \quad -\mathbf{v} \cdot \mathbf{g} = T(\tfrac{1}{2}g^2 - (\boldsymbol{\omega} \times \mathbf{v}) \cdot \mathbf{g}),$$

$$\Rightarrow \quad T = -\frac{\mathbf{v} \cdot \mathbf{g}}{\tfrac{1}{2}g^2}\left[1 - \frac{(\boldsymbol{\omega} \times \mathbf{v}) \cdot \mathbf{g}}{\tfrac{1}{2}g^2}\right]^{-1}$$

$$\approx \tau\left[1 + \frac{2(\boldsymbol{\omega} \times \mathbf{v}) \cdot \mathbf{g}}{g^2} + \cdots\right].$$

Working to first order in ω, we may put $T = \tau$ in those terms in $(*)$ that involve another factor ω, namely $\boldsymbol{\omega} \times \mathbf{v}$ and $\boldsymbol{\omega} \times \mathbf{g}$. We then find, to this order, that

$$\mathbf{s} + \boldsymbol{\Delta} = \frac{1}{2}\mathbf{g}\left[\tau^2 + \frac{4(\boldsymbol{\omega} \times \mathbf{v}) \cdot \mathbf{g}}{g^2}\tau^2 + \cdots\right] - \frac{1}{3}(\boldsymbol{\omega} \times \mathbf{g})\tau^3$$

$$- (\boldsymbol{\omega} \times \mathbf{v})\tau^2 + \mathbf{v}\tau + 2\frac{(\boldsymbol{\omega} \times \mathbf{v}) \cdot \mathbf{g}}{g^2}\mathbf{v}\tau$$

$$= \mathbf{s} + \frac{(\boldsymbol{\omega} \times \mathbf{v}) \cdot \mathbf{g}}{g^2}(2\mathbf{g}\tau^2 + 2\mathbf{v}\tau) - \frac{1}{3}(\boldsymbol{\omega} \times \mathbf{g})\tau^3 - (\boldsymbol{\omega} \times \mathbf{v})\tau^2.$$

Hence, as stated in the question,

$$\Delta = \frac{2\tau}{g^2}[(\mathbf{g} \times \boldsymbol{\omega}) \cdot \mathbf{v}](g\tau + v) - (\boldsymbol{\omega} \times \mathbf{v})\tau^2 - \frac{1}{3}(\boldsymbol{\omega} \times \mathbf{g})\tau^3.$$

(c) With the ship at latitude $45°$ and firing the shell at close to $45°$ to the local horizontal, $\mathbf{v}$ and $\boldsymbol{\omega}$ are almost parallel and the $\boldsymbol{\omega} \times \mathbf{v}$ term can be set to zero. Further, with $\mathbf{v}$ in a northerly direction, $(\mathbf{g} \times \boldsymbol{\omega}) \cdot \mathbf{v} = 0$.

Thus we are left with only the cubic term in τ. In this,

$$\tau = \frac{2 \times 300 \cos(\pi/4)}{9.8} = 43.3 \text{ s},$$

and $\boldsymbol{\omega} \times \mathbf{g}$ is in a westerly direction (recall that $\boldsymbol{\omega}$ is directed northwards and $\mathbf{g}$ is directed downwards, towards the origin) and of magnitude $7 \times 10^{-5} \times 9.8 \sin(\pi/4) = 4.85 \times 10^{-4}$ m s^{-3}. Thus the miss is by approximately

$$-\tfrac{1}{3} \times 4.85 \times 10^{-4} \times (43.3)^3 = -13 \text{ m},$$

i.e. some 10–15 m to the East of its intended target.

11.9 Parameterising the hyperboloid

$$\frac{x^2}{a^2} + \frac{y^2}{b^2} - \frac{z^2}{c^2} = 1$$

by $x = a \cos\theta \sec\phi$, $y = b \sin\theta \sec\phi$, $z = c \tan\phi$, show that an area element on its surface is

$$dS = \sec^2\phi \left[c^2 \sec^2\phi \left(b^2 \cos^2\theta + a^2 \sin^2\theta\right) + a^2 b^2 \tan^2\phi\right]^{1/2} d\theta \, d\phi.$$

Use this formula to show that the area of the curved surface $x^2 + y^2 - z^2 = a^2$ between the planes $z = 0$ and $z = 2a$ is

$$\pi a^2 \left(6 + \frac{1}{\sqrt{2}} \sinh^{-1} 2\sqrt{2}\right).$$

With $x = a \cos\theta \sec\phi$, $y = b \sin\theta \sec\phi$ and $z = c \tan\phi$, the tangent vectors to the surface are given in Cartesian coordinates by

$$\frac{d\mathbf{r}}{d\theta} = (-a \sin\theta \sec\phi, \; b \cos\theta \sec\phi, \; 0),$$

$$\frac{d\mathbf{r}}{d\phi} = (a \cos\theta \sec\phi \tan\phi, \; b \sin\theta \sec\phi \tan\phi, \; c \sec^2\phi),$$

and the element of area by

$$dS = \left|\frac{d\mathbf{r}}{d\theta} \times \frac{d\mathbf{r}}{d\phi}\right| d\theta \, d\phi$$

$$= \left|(bc \cos\theta \sec^3\phi, \; ac \sin\theta \sec^3\phi, \; -ab \sec^2\phi \tan\phi)\right| d\theta \, d\phi$$

$$= \sec^2\phi \left[c^2 \sec^2\phi \left(b^2 \cos^2\theta + a^2 \sin^2\theta\right) + a^2 b^2 \tan^2\phi\right]^{1/2} d\theta \, d\phi.$$

We set $b = c = a$ and note that the plane $z = 2a$ corresponds to $\phi = \tan^{-1} 2$. The ranges of integration are therefore $0 \le \theta < 2\pi$ and $0 \le \phi \le \tan^{-1} 2$, whilst

$$dS = \sec^2 \phi (a^4 \sec^2 \phi + a^4 \tan^2 \phi)^{1/2} \, d\theta \, d\phi,$$

i.e. it is independent of θ.

To evaluate the integral of dS, we set $\tan \phi = \sinh \psi / \sqrt{2}$, with

$$\sec^2 \phi \, d\phi = \frac{1}{\sqrt{2}} \cosh \psi \, d\psi \quad \text{and} \quad \sec^2 \phi = 1 + \tfrac{1}{2} \sinh^2 \psi.$$

The upper limit for ψ will be given by $\Psi = \sinh^{-1} 2\sqrt{2}$; we note that $\cosh \Psi = 3$. Integrating over θ and making the above substitutions yields

$$S = 2\pi \int_0^\Psi \frac{1}{\sqrt{2}} \cosh \psi \, d\psi \, a^2 \left(1 + \frac{1}{2} \sinh^2 \psi + \frac{1}{2} \sinh^2 \psi \right)^{1/2}$$

$$= \sqrt{2}\pi a^2 \int_0^\Psi \cosh^2 \psi \, d\psi$$

$$= \frac{\sqrt{2}\pi a^2}{2} \int_0^\Psi (\cosh 2\psi + 1) \, d\psi$$

$$= \frac{\sqrt{2}\pi a^2}{2} \left[\frac{\sinh 2\psi}{2} + \psi \right]_0^\Psi$$

$$= \frac{\pi a^2}{\sqrt{2}} [\sinh \psi \cosh \psi + \psi]_0^\Psi$$

$$= \frac{\pi a^2}{\sqrt{2}} [(2\sqrt{2})(3) + \sinh^{-1} 2\sqrt{2}] = \pi a^2 \left(6 + \frac{1}{\sqrt{2}} \sinh^{-1} 2\sqrt{2} \right).$$

11.11 Verify by direct calculation that

$$\nabla \cdot (\mathbf{a} \times \mathbf{b}) = \mathbf{b} \cdot (\nabla \times \mathbf{a}) - \mathbf{a} \cdot (\nabla \times \mathbf{b}).$$

The proof of this standard result for the divergence of a vector product is most easily carried out in Cartesian coordinates though, of course, the result is valid in any three-dimensional coordinate system.

$$\text{LHS} = \nabla \cdot (\mathbf{a} \times \mathbf{b})$$

$$= \frac{\partial}{\partial x}(a_y b_z - a_z b_y) + \frac{\partial}{\partial y}(a_z b_x - a_x b_z) + \frac{\partial}{\partial z}(a_x b_y - a_y b_x)$$

$$= a_x \left(-\frac{\partial b_z}{\partial y} + \frac{\partial b_y}{\partial z} \right) + a_y \left(\frac{\partial b_z}{\partial x} - \frac{\partial b_x}{\partial z} \right) + a_z \left(-\frac{\partial b_y}{\partial x} + \frac{\partial b_x}{\partial y} \right)$$

$$+ b_x \left(\frac{\partial a_z}{\partial y} - \frac{\partial a_y}{\partial z} \right) + b_y \left(-\frac{\partial a_z}{\partial x} + \frac{\partial a_x}{\partial z} \right) + b_z \left(\frac{\partial a_y}{\partial x} - \frac{\partial a_x}{\partial y} \right)$$

$$= -\mathbf{a} \cdot (\nabla \times \mathbf{b}) + \mathbf{b} \cdot (\nabla \times \mathbf{a}) = \text{RHS}.$$

11.13 Evaluate the Laplacian of the function

$$\psi(x, y, z) = \frac{zx^2}{x^2 + y^2 + z^2}$$

(a) directly in Cartesian coordinates and (b) after changing to a spherical polar coordinate system. Verify that, as they must, the two methods give the same result.

(a) In Cartesian coordinates we need to evaluate

$$\nabla^2 \psi = \frac{\partial^2 \psi}{\partial x^2} + \frac{\partial^2 \psi}{\partial y^2} + \frac{\partial^2 \psi}{\partial z^2}.$$

The required derivatives are

$$\frac{\partial \psi}{\partial x} = \frac{2xz(y^2 + z^2)}{(x^2 + y^2 + z^2)^2}, \quad \frac{\partial^2 \psi}{\partial x^2} = \frac{(y^2 + z^2)(2zy^2 + 2z^3 - 6x^2z)}{(x^2 + y^2 + z^2)^3},$$

$$\frac{\partial \psi}{\partial y} = \frac{-2x^2yz}{(x^2 + y^2 + z^2)^2}, \quad \frac{\partial^2 \psi}{\partial y^2} = -\frac{2zx^2(x^2 + z^2 - 3y^2)}{(x^2 + y^2 + z^2)^3},$$

$$\frac{\partial \psi}{\partial z} = \frac{x^2(x^2 + y^2 - z^2)}{(x^2 + y^2 + z^2)^2}, \quad \frac{\partial^2 \psi}{\partial z^2} = -\frac{2zx^2(3x^2 + 3y^2 - z^2)}{(x^2 + y^2 + z^2)^3}.$$

Thus, writing $r^2 = x^2 + y^2 + z^2$,

$$\nabla^2 \psi = \frac{2z[\,(y^2 + z^2)(y^2 + z^2 - 3x^2) - 4x^4\,]}{(x^2 + y^2 + z^2)^3}$$

$$= \frac{2z[\,(r^2 - x^2)(r^2 - 4x^2) - 4x^4\,]}{r^6}$$

$$= \frac{2z(r^2 - 5x^2)}{r^4}.$$

(b) In spherical polar coordinates,

$$\psi(r, \theta, \phi) = \frac{r \cos\theta\, r^2 \sin^2\theta \cos^2\phi}{r^2} = r \cos\theta \sin^2\theta \cos^2\phi.$$

The three contributions to $\nabla^2 \psi$ in spherical polars are

$$(\nabla^2 \psi)_r = \frac{1}{r^2} \frac{\partial}{\partial r} \left(r^2 \frac{\partial \psi}{\partial r} \right)$$

$$= \frac{2}{r} \cos\theta \sin^2\theta \cos^2\phi,$$

$$(\nabla^2 \psi)_\theta = \frac{1}{r^2 \sin\theta} \frac{\partial}{\partial \theta} \left(\sin\theta \frac{\partial \psi}{\partial \theta} \right)$$

$$= \frac{1}{r} \frac{\cos^2\phi}{\sin\theta} \frac{\partial}{\partial \theta} \left[\sin\theta \frac{\partial}{\partial \theta} (\cos\theta \sin^2\theta) \right]$$

$$= \frac{\cos^2\phi}{r} (4 \cos^3\theta - 8 \sin^2\theta \cos\theta),$$

$$(\nabla^2 \psi)_\phi = \frac{1}{r^2 \sin^2 \theta} \frac{\partial^2 \psi}{\partial \phi^2}$$

$$= \frac{\cos \theta}{r}(-2\cos^2 \phi + 2\sin^2 \phi).$$

Thus, the full Laplacian in spherical polar coordinates reads

$$\nabla^2 \psi = \frac{\cos \theta}{r}(2\sin^2 \theta \cos^2 \phi + 4\cos^2 \theta \cos^2 \phi - 8\sin^2 \theta \cos^2 \phi - 2\cos^2 \phi + 2\sin^2 \phi)$$

$$= \frac{\cos \theta}{r}(4\cos^2 \phi - 10\sin^2 \theta \cos^2 \phi - 2\cos^2 \phi + 2\sin^2 \phi)$$

$$= \frac{\cos \theta}{r}(2 - 10\sin^2 \theta \cos^2 \phi)$$

$$= \frac{2r\cos \theta(r^2 - 5r^2 \sin^2 \theta \cos^2 \phi)}{r^4}.$$

Rewriting this last expression in terms of Cartesian coordinates, one finally obtains

$$\nabla^2 \psi = \frac{2z(r^2 - 5x^2)}{r^4},$$

which establishes the equivalence of the two approaches.

11.15 The (Maxwell) relationship between a time-independent magnetic field $\mathbf{B}$ and the current density $\mathbf{J}$ (measured in SI units in $\mathrm{A\,m^{-2}}$) producing it,

$$\nabla \times \mathbf{B} = \mu_0 \mathbf{J},$$

can be applied to a long cylinder of conducting ionised gas which, in cylindrical polar coordinates, occupies the region $\rho < a$.

(a) Show that a uniform current density $(0, C, 0)$ and a magnetic field $(0, 0, B)$, with B constant ($= B_0$) for $\rho > a$ and $B = B(\rho)$ for $\rho < a$, are consistent with this equation. Given that $B(0) = 0$ and that $\mathbf{B}$ is continuous at $\rho = a$, obtain expressions for C and $B(\rho)$ in terms of B_0 and a.

(b) The magnetic field can be expressed as $\mathbf{B} = \nabla \times \mathbf{A}$, where $\mathbf{A}$ is known as the vector potential. Show that a suitable $\mathbf{A}$ can be found which has only one non-vanishing component, $A_\phi(\rho)$, and obtain explicit expressions for $A_\phi(\rho)$ for both $\rho < a$ and $\rho > a$. Like $\mathbf{B}$, the vector potential is continuous at $\rho = a$.

(c) The gas pressure $p(\rho)$ satisfies the hydrostatic equation $\nabla p = \mathbf{J} \times \mathbf{B}$ and vanishes at the outer wall of the cylinder. Find a general expression for p.

(a) In cylindrical polars with $\mathbf{B} = (0, 0, B(\rho))$, for $\rho \leq a$ we have

$$\mu_0(0, \ C, \ 0) = \nabla \times \mathbf{B} = \left(\frac{1}{\rho}\frac{\partial B}{\partial \phi}, \ -\frac{\partial B}{\partial \rho}, \ 0\right).$$

As expected, $\partial B/\partial\phi = 0$. The azimuthal component of the equation gives

$$-\frac{\partial B}{\partial \rho} = \mu_0 C \quad \text{for} \quad \rho \leq a \quad \Rightarrow \quad B(\rho) = B(0) - \mu_0 C \rho.$$

Since **B** has to be differentiable at the origin of ρ and have no ϕ-dependence, $B(0)$ must be zero. This, together with $B = B_0$ for $\rho > a$, requires that $C = -B_0/(a\mu_0)$ and $B(\rho) = B_0\rho/a$ for $0 \leq \rho \leq a$.

(b) With $\mathbf{B} = \nabla \times \mathbf{A}$, consider **A** of the form $\mathbf{A} = (0, \ A(\rho), \ 0)$. Then

$$(0, \ 0, \ B(\rho)) = \frac{1}{\rho}\left(\frac{\partial}{\partial z}(\rho A), \ 0, \ \frac{\partial}{\partial \rho}(\rho A)\right)$$

$$= \left(0, \ 0, \ \frac{1}{\rho}\frac{\partial}{\partial \rho}(\rho A)\right).$$

We now equate the only non-vanishing component on each side of the above equation, treating inside and outside the cylinder separately.

For $0 < \rho \leq a$,

$$\frac{1}{\rho}\frac{\partial}{\partial \rho}(\rho A) = \frac{B_0\rho}{a},$$

$$\rho A = \frac{B_0\rho^3}{3a} + D,$$

$$A(\rho) = \frac{B_0\rho^2}{3a} + \frac{D}{\rho}.$$

Since $A(0)$ must be finite (so that A is differentiable there), $D = 0$.

For $\rho > a$,

$$\frac{1}{\rho}\frac{\partial}{\partial \rho}(\rho A) = B_0,$$

$$\rho A = \frac{B_0\rho^2}{2} + E,$$

$$A(\rho) = \frac{1}{2}B_0\rho + \frac{E}{\rho}.$$

At $\rho = a$, the continuity of **A** requires

$$\frac{B_0 a^2}{3a} = \frac{1}{2}B_0 a + \frac{E}{a} \quad \Rightarrow \quad E = -\frac{B_0 a^2}{6}.$$

Thus, to summarise,

$$A(\rho) = \frac{B_0\rho^2}{3a} \quad \text{for} \quad 0 \leq \rho \leq a,$$

and

$$A(\rho) = B_0\left(\frac{\rho}{2} - \frac{a^2}{6\rho}\right) \quad \text{for} \quad \rho \geq a.$$

(c) For the gas pressure $p(\rho)$ in the region $0 < \rho \le a$, we have $\nabla p = \mathbf{J} \times \mathbf{B}$. In component form,

$$\left(\frac{dp}{d\rho},\ 0,\ 0 \right) = \left(0,\ -\frac{B_0}{a\mu_0},\ 0 \right) \times \left(0,\ 0,\ \frac{B_0\rho}{a} \right),$$

with $p(a) = 0$.

$$\frac{dp}{d\rho} = -\frac{B_0^2\rho}{\mu_0 a^2} \quad \Rightarrow \quad p(\rho) = \frac{B_0^2}{2\mu_0}\left[1 - \left(\frac{\rho}{a}\right)^2 \right].$$

11.17 Maxwell's equations for electromagnetism in free space (i.e. in the absence of charges, currents and dielectric or magnetic media) can be written

(i) $\nabla \cdot \mathbf{B} = 0$,   (ii) $\nabla \cdot \mathbf{E} = 0$,

(iii) $\nabla \times \mathbf{E} + \dfrac{\partial \mathbf{B}}{\partial t} = \mathbf{0}$,   (iv) $\nabla \times \mathbf{B} - \dfrac{1}{c^2}\dfrac{\partial \mathbf{E}}{\partial t} = \mathbf{0}$.

A vector $\mathbf{A}$ is defined by $\mathbf{B} = \nabla \times \mathbf{A}$, and a scalar ϕ by $\mathbf{E} = -\nabla\phi - \partial\mathbf{A}/\partial t$. Show that if the condition

$$\text{(v)} \quad \nabla \cdot \mathbf{A} + \frac{1}{c^2}\frac{\partial\phi}{\partial t} = 0$$

is imposed (this is known as choosing the Lorentz gauge), then $\mathbf{A}$ and ϕ satisfy wave equations as follows:

$$\text{(vi)} \quad \nabla^2\phi - \frac{1}{c^2}\frac{\partial^2\phi}{\partial t^2} = 0,$$

$$\text{(vii)} \quad \nabla^2\mathbf{A} - \frac{1}{c^2}\frac{\partial^2\mathbf{A}}{\partial t^2} = \mathbf{0}.$$

The reader is invited to proceed as follows.

(a) Verify that the expressions for $\mathbf{B}$ and $\mathbf{E}$ in terms of $\mathbf{A}$ and ϕ are consistent with (i) and (iii).
(b) Substitute for $\mathbf{E}$ in (ii) and use the derivative with respect to time of (v) to eliminate $\mathbf{A}$ from the resulting expression. Hence obtain (vi).
(c) Substitute for $\mathbf{B}$ and $\mathbf{E}$ in (iv) in terms of $\mathbf{A}$ and ϕ. Then use the gradient of (v) to simplify the resulting equation and so obtain (vii).

(a) Substituting for $\mathbf{B}$ in (i),

$$\nabla \cdot \mathbf{B} = \nabla \cdot (\nabla \times \mathbf{A}) = 0, \quad \text{as it is for } any \text{ vector } \mathbf{A}.$$

Substituting for $\mathbf{E}$ and $\mathbf{B}$ in (iii),

$$\nabla \times \mathbf{E} + \frac{\partial \mathbf{B}}{\partial t} = -(\nabla \times \nabla\phi) - \nabla \times \frac{\partial\mathbf{A}}{\partial t} + \frac{\partial}{\partial t}(\nabla \times \mathbf{A}) = \mathbf{0}.$$

Here we have used the facts that $\nabla \times \nabla\phi = \mathbf{0}$ for any scalar and that, since $\partial/\partial t$ and ∇ act on different variables, the order in which they are applied to $\mathbf{A}$ can be reversed. Thus (i) and (iii) are automatically satisfied if $\mathbf{E}$ and $\mathbf{B}$ are represented in terms of $\mathbf{A}$ and ϕ.

(b) Substituting for $\mathbf{E}$ in (ii) and taking the time derivative of (v),

$$0 = \nabla \cdot \mathbf{E} = -\nabla^2 \phi - \frac{\partial}{\partial t} (\nabla \cdot \mathbf{A}),$$

$$0 = \frac{\partial}{\partial t} (\nabla \cdot \mathbf{A}) + \frac{1}{c^2} \frac{\partial^2 \phi}{\partial t^2}.$$

Adding these equations gives

$$0 = -\nabla^2 \phi + \frac{1}{c^2} \frac{\partial^2 \phi}{\partial t^2}.$$

This is result (vi), the wave equation for ϕ.

(c) Substituting for $\mathbf{B}$ and $\mathbf{E}$ in (iv) and taking the gradient of (v),

$$\nabla \times (\nabla \times \mathbf{A}) - \frac{1}{c^2} \left(-\frac{\partial}{\partial t} \nabla \phi - \frac{\partial^2 \mathbf{A}}{\partial t^2} \right) = 0,$$

$$\nabla(\nabla \cdot \mathbf{A}) - \nabla^2 \mathbf{A} + \frac{1}{c^2} \frac{\partial}{\partial t} (\nabla \phi) + \frac{1}{c^2} \frac{\partial^2 \mathbf{A}}{\partial t^2} = 0.$$

From (v),

$$\nabla(\nabla \cdot \mathbf{A}) + \frac{1}{c^2} \frac{\partial}{\partial t} (\nabla \phi) = 0.$$

Subtracting these gives

$$-\nabla^2 \mathbf{A} + \frac{1}{c^2} \frac{\partial^2 \mathbf{A}}{\partial t^2} = 0.$$

In the second line we have used the vector identity

$$\nabla^2 \mathbf{F} = \nabla(\nabla \cdot \mathbf{F}) - \nabla \times (\nabla \times \mathbf{F})$$

to replace $\nabla \times (\nabla \times \mathbf{A})$. The final equation is result (vii).

11.19 Paraboloidal coordinates u, v, ϕ are defined in terms of Cartesian coordinates by

$$x = uv \cos \phi, \qquad y = uv \sin \phi, \qquad z = \tfrac{1}{2}(u^2 - v^2).$$

Identify the coordinate surfaces in the u, v, ϕ system. Verify that each coordinate surface ($u =$ constant, say) intersects every coordinate surface on which one of the other two coordinates (v, say) is constant. Show further that the system of coordinates is an orthogonal one and determine its scale factors. Prove that the u-component of $\nabla \times \mathbf{a}$ is given by

$$\frac{1}{(u^2 + v^2)^{1/2}} \left(\frac{a_\phi}{v} + \frac{\partial a_\phi}{\partial v} \right) - \frac{1}{uv} \frac{\partial a_v}{\partial \phi}.$$

To find a surface of constant u we eliminate v from the given relationships:

$$x^2 + y^2 = u^2 v^2 \quad \Rightarrow \quad 2z = u^2 - \frac{x^2 + y^2}{u^2}.$$

This is an inverted paraboloid of revolution about the z-axis. The range of z is $-\infty < z \le \tfrac{1}{2}u^2$.

Similarly, the surface of constant v is given by

$$2z = \frac{x^2 + y^2}{v^2} - v^2.$$

This is also a paraboloid of revolution about the z-axis, but this time it is not inverted. The range of z is $-\frac{1}{2}v^2 \le z < \infty$.

Since every constant-u paraboloid has some part of its surface in the region $z > 0$ and every constant-v paraboloid has some part of its surface in the region $z < 0$, it follows that every member of the first set intersects each member of the second, and vice versa.

The surfaces of constant ϕ, $y = x \tan \phi$, are clearly (half-) planes containing the z-axis; each cuts the members of the other two sets in parabolic lines.

We now determine (the Cartesian components of) the tangential vectors and test their orthogonality:

$$\mathbf{e}_1 = \frac{\partial \mathbf{r}}{\partial u} = (v \cos \phi, \ v \sin \phi, \ u),$$

$$\mathbf{e}_2 = \frac{\partial \mathbf{r}}{\partial v} = (u \cos \phi, \ u \sin \phi, \ -v),$$

$$\mathbf{e}_3 = \frac{\partial \mathbf{r}}{\partial \phi} = (-uv \sin \phi, \ uv \cos \phi, \ 0),$$

$$\mathbf{e}_1 \cdot \mathbf{e}_2 = uv(\cos \phi \cos \phi + \sin \phi \sin \phi) - uv = 0,$$

$$\mathbf{e}_2 \cdot \mathbf{e}_3 = u^2 v(-\cos \phi \sin \phi + \sin \phi \cos \phi) = 0,$$

$$\mathbf{e}_1 \cdot \mathbf{e}_3 = uv^2(-\cos \phi \sin \phi + \sin \phi \cos \phi) = 0.$$

This shows that all pairs of tangential vectors are orthogonal and therefore that the coordinate system is an orthogonal one. Its scale factors are given by the magnitudes of these tangential vectors:

$$h_u^2 = |\mathbf{e}_1|^2 = (v \cos \phi)^2 + (v \sin \phi)^2 + u^2 = u^2 + v^2,$$

$$h_v^2 = |\mathbf{e}_2|^2 = (u \cos \phi)^2 + (u \sin \phi)^2 + v^2 = u^2 + v^2,$$

$$h_\phi^2 = |\mathbf{e}_3|^2 = (uv \sin \phi)^2 + (uv \cos \phi)^2 = u^2 v^2.$$

Thus

$$h_u = h_v = \sqrt{u^2 + v^2}, \qquad h_\phi = uv.$$

The u-component of $\nabla \times \mathbf{a}$ is given by

$$[\nabla \times \mathbf{a}]_u = \frac{h_u}{h_u h_v h_\phi} \left[\frac{\partial}{\partial v} (h_\phi a_\phi) - \frac{\partial}{\partial \phi} (h_v a_v) \right]$$

$$= \frac{1}{uv\sqrt{u^2 + v^2}} \left[\frac{\partial}{\partial v} (uv a_\phi) - \frac{\partial}{\partial \phi} (\sqrt{u^2 + v^2} \, a_v) \right]$$

$$= \frac{1}{\sqrt{u^2 + v^2}} \left(\frac{a_\phi}{v} + \frac{\partial a_\phi}{\partial v} \right) - \frac{1}{uv} \frac{\partial a_v}{\partial \phi},$$

as stated in the question.

11.21 Hyperbolic coordinates u, v, ϕ are defined in terms of Cartesian coordinates by

$$x = \cosh u \cos v \cos \phi, \qquad y = \cosh u \cos v \sin \phi, \qquad z = \sinh u \sin v.$$

Sketch the coordinate curves in the $\phi = 0$ plane, showing that far from the origin they become concentric circles and radial lines. In particular, identify the curves $u = 0$, $v = 0$, $v = \pi/2$ and $v = \pi$. Calculate the tangent vectors at a general point, show that they are mutually orthogonal and deduce that the appropriate scale factors are

$$h_u = h_v = (\cosh^2 u - \cos^2 v)^{1/2}, \qquad h_\phi = \cosh u \cos v.$$

Find the most general function $\psi(u)$ of u only that satisfies Laplace's equation $\nabla^2 \psi = 0$.

In the plane $\phi = 0$, i.e. $y = 0$, the curves $u = $ constant have x and z connected by

$$\frac{x^2}{\cosh^2 u} + \frac{z^2}{\sinh^2 u} = 1.$$

This general form is that of an ellipse, with foci at $(\pm 1, 0)$. With $u = 0$, it is the line joining the two foci (covered twice). As $u \to \infty$, and $\cosh u \approx \sinh u$, the form becomes that of a circle of very large radius.

The curves $v = $ constant are expressed by

$$\frac{x^2}{\cos^2 v} - \frac{z^2}{\sin^2 v} = 1.$$

These curves are hyperbolae that, for large x and z and fixed v, approximate $z = \pm x \tan v$, i.e. radial lines. The curve $v = 0$ is the part of the x-axis $1 \le x \le \infty$ (covered twice), whilst the curve $v = \pi$ is its reflection in the z-axis. The curve $v = \pi/2$ *is the z-axis*.

In Cartesian coordinates a general point and its derivatives with respect to u, v and ϕ are given by

$$\mathbf{r} = \cosh u \cos v \cos \phi\, \mathbf{i} + \cosh u \cos v \sin \phi\, \mathbf{j} + \sinh u \sin v\, \mathbf{k},$$

$$\mathbf{e}_1 = \frac{\partial \mathbf{r}}{\partial u} = \sinh u \cos v \cos \phi\, \mathbf{i} + \sinh u \cos v \sin \phi\, \mathbf{j} + \cosh u \sin v\, \mathbf{k},$$

$$\mathbf{e}_2 = \frac{\partial \mathbf{r}}{\partial v} = -\cosh u \sin v \cos \phi\, \mathbf{i} - \cosh u \sin v \sin \phi\, \mathbf{j} + \sinh u \cos v\, \mathbf{k},$$

$$\mathbf{e}_3 = \frac{\partial \mathbf{r}}{\partial \phi} = \cosh u \cos v(-\sin \phi\, \mathbf{i} + \cos \phi\, \mathbf{j}).$$

Now consider the scalar products:

$$\mathbf{e}_1 \cdot \mathbf{e}_2 = \sinh u \cos v \cosh u \sin v(-\cos^2 \phi - \sin^2 \phi + 1) = 0,$$

$$\mathbf{e}_1 \cdot \mathbf{e}_3 = \sinh u \cos^2 v \cosh u(-\sin \phi \cos \phi + \sin \phi \cos \phi) = 0,$$

$$\mathbf{e}_2 \cdot \mathbf{e}_3 = \cosh^2 u \sin v \cos v(\sin \phi \cos \phi - \sin \phi \cos \phi) = 0.$$

As each is zero, the system is an orthogonal one.

The scale factors are given by $|\mathbf{e}_i|$ and are thus found from:

$$|\mathbf{e}_1|^2 = \sinh^2 u \cos^2 v (\cos^2 \phi + \sin^2 \phi) + \cosh^2 u \sin^2 v$$

$$= (\cosh^2 u - 1) \cos^2 v + \cosh^2 u (1 - \cos^2 v)$$

$$= \cosh^2 u - \cos^2 v;$$

$$|\mathbf{e}_2|^2 = \cosh^2 u \sin^2 v (\cos^2 \phi + \sin^2 \phi) + \sinh^2 u \cos^2 v$$

$$= \cosh^2 u (1 - \cos^2 v) + (\cosh^2 u - 1) \cos^2 v$$

$$= \cosh^2 u - \cos^2 v;$$

$$|\mathbf{e}_3|^2 = \cosh^2 u \cos^2 v (\sin^2 \phi + \cos^2 \phi) = \cosh^2 u \cos^2 v.$$

The immediate deduction is that

$$h_u = h_v = (\cosh^2 u - \cos^2 v)^{1/2}, \qquad h_\phi = \cosh u \cos v.$$

An alternative form for h_u and h_v is $(\sinh^2 u + \sin^2 v)^{1/2}$.

If a solution of Laplace's equation is to be a function, $\psi(u)$, of u only, then all differentiation with respect to v and ϕ can be ignored. The expression for $\nabla^2 \psi$ reduces to

$$\nabla^2 \psi = \frac{1}{h_u h_v h_\phi} \left[\frac{\partial}{\partial u} \left(\frac{h_v h_\phi}{h_u} \frac{\partial \psi}{\partial u} \right) \right]$$

$$= \frac{1}{\cosh u \cos v (\cosh^2 u - \cos^2 v)} \left[\frac{\partial}{\partial u} \left(\cosh u \cos v \frac{\partial \psi}{\partial u} \right) \right].$$

Laplace's equation itself is even simpler and reduces to

$$\frac{\partial}{\partial u} \left(\cosh u \frac{\partial \psi}{\partial u} \right) = 0.$$

This can be rewritten as

$$\frac{\partial \psi}{\partial u} = \frac{k}{\cosh u} = \frac{2k}{e^u + e^{-u}} = \frac{2k e^u}{e^{2u} + 1},$$

$$d\psi = \frac{A e^u \, du}{1 + (e^u)^2} \quad \Rightarrow \quad \psi = B \tan^{-1} e^u + c.$$

This is the most general function of u only that satisfies Laplace's equation.

12 Line, surface and volume integrals

12.1 The vector field $\mathbf{F}$ is defined by

$$\mathbf{F} = 2xz\mathbf{i} + 2yz^2\mathbf{j} + (x^2 + 2y^2z - 1)\mathbf{k}.$$

Calculate $\nabla \times \mathbf{F}$ and deduce that $\mathbf{F}$ can be written $F = \nabla\phi$. Determine the form of ϕ.

With $\mathbf{F}$ as given, we calculate the curl of $\mathbf{F}$ to see whether or not it is the zero vector:

$$\nabla \times \mathbf{F} = (4yz - 4yz, \; 2x - 2x, \; 0 - 0) = \mathbf{0}.$$

The fact that it is implies that $\mathbf{F}$ can be written as $\nabla\phi$ for some scalar ϕ.

The form of $\phi(x, y, z)$ is found by integrating, in turn, the components of $\mathbf{F}$ until consistency is achieved, i.e. until a ϕ is found that has partial derivatives equal to the corresponding components of $\mathbf{F}$:

$$2xz = F_x = \frac{\partial\phi}{\partial x} \quad \Rightarrow \quad \phi(x, y, z) = x^2z + g(y, z),$$

$$2yz^2 = F_y = \frac{\partial}{\partial y}[x^2z + g(y, z)] \quad \Rightarrow \quad g(y, z) = y^2z^2 + h(z),$$

$$x^2 + 2y^2z - 1 = F_z = \frac{\partial}{\partial z}[x^2z + y^2z^2 + h(z)]$$

$$\Rightarrow \quad h(z) = -z + k.$$

Hence, to within an unimportant constant, the form of ϕ is

$$\phi(x, y, z) = x^2z + y^2z^2 - z.$$

12.3 A vector field $\mathbf{F}$ is given by $\mathbf{F} = xy^2\mathbf{i} + 2\mathbf{j} + x\mathbf{k}$ and L is a path parameterised by $x = ct$, $y = c/t$, $z = d$ for the range $1 \leq t \leq 2$. Evaluate the three integrals

(a) $\displaystyle\int_L \mathbf{F}\, dt,$ (b) $\displaystyle\int_L \mathbf{F}\, dy,$ (c) $\displaystyle\int_L \mathbf{F} \cdot d\mathbf{r}.$

Although all three integrals are along the same path L, they are not necessarily of the same type. The vector or scalar nature of the integral is determined by that of the integrand when it is expressed in a form containing the infinitesimal dt.

(a) This is a vector integral and contains three separate integrations. We express each of the integrands in terms of t, according to the parameterisation of the integration path

L, before integrating:

$$\int_L \mathbf{F}\,dt = \int_1^2 \left(\frac{c^3}{t}\mathbf{i} + 2\mathbf{j} + ct\,\mathbf{k}\right) dt$$

$$= \left[c^3 \ln t\,\mathbf{i} + 2t\,\mathbf{j} + \frac{1}{2}ct^2\,\mathbf{k}\right]_1^2$$

$$= c^3 \ln 2\,\mathbf{i} + 2\mathbf{j} + \frac{3}{2}c\,\mathbf{k}.$$

(b) This is a similar vector integral but here we must also replace the infinitesimal dy by the infinitesimal $-c\,dt/t^2$ before integrating:

$$\int_L \mathbf{F}\,dy = \int_1^2 \left(\frac{c^3}{t}\mathbf{i} + 2\mathbf{j} + ct\,\mathbf{k}\right)\left(\frac{-c}{t^2}\right) dt$$

$$= \left[\frac{c^4}{2t^2}\mathbf{i} + \frac{2c}{t}\mathbf{j} - c^2 \ln t\,\mathbf{k}\right]_1^2$$

$$= -\frac{3c^4}{8}\mathbf{i} - c\mathbf{j} - c^2 \ln 2\,\mathbf{k}.$$

(c) This is a scalar integral and before integrating we must take the scalar product of $\mathbf{F}$ with $d\mathbf{r} = dx\,\mathbf{i} + dy\,\mathbf{j} + dz\,\mathbf{k}$ to give a single integrand:

$$\int_L \mathbf{F}\cdot d\mathbf{r} = \int_1^2 \left(\frac{c^3}{t}\mathbf{i} + 2\mathbf{j} + ct\,\mathbf{k}\right) \cdot \left(c\,\mathbf{i} - \frac{c}{t^2}\mathbf{j} + 0\mathbf{k}\right) dt$$

$$= \int_1^2 \left(\frac{c^4}{t} - \frac{2c}{t^2}\right) dt$$

$$= \left[c^4 \ln t + \frac{2c}{t}\right]_1^2$$

$$= c^4 \ln 2 - c.$$

12.5 Determine the point of intersection P, in the first quadrant, of the two ellipses

$$\frac{x^2}{a^2} + \frac{y^2}{b^2} = 1 \quad \text{and} \quad \frac{x^2}{b^2} + \frac{y^2}{a^2} = 1.$$

Taking $b < a$, consider the contour L that bounds the area in the first quadrant that is common to the two ellipses. Show that the parts of L that lie along the coordinate axes contribute nothing to the line integral around L of $x\,dy - y\,dx$. Using a parameterisation of each ellipse of the general form $x = X\cos\phi$ and $y = Y\sin\phi$, evaluate the two remaining line integrals and hence find the total area common to the two ellipses.

Note: the line integral of $x\,dy - y\,dx$ around a general closed convex contour is equal to twice the area enclosed by that contour.

From the symmetry of the equations under the interchange of x and y, the point P must have $x = y$. Thus,

$$x^2 \left(\frac{1}{a^2} + \frac{1}{b^2} \right) = 1 \quad \Rightarrow \quad x = \frac{ab}{(a^2 + b^2)^{1/2}}.$$

Denoting as curve C_1 the part of

$$\frac{x^2}{a^2} + \frac{y^2}{b^2} = 1$$

that lies on the boundary of the common region, we parameterise it by $x = a \cos \theta_1$ and $y = b \sin \theta_1$. Curve C_1 starts from P and finishes on the y-axis. At P,

$$a \cos \theta_1 = x = \frac{ab}{(a^2 + b^2)^{1/2}} \quad \Rightarrow \quad \tan \theta_1 = \frac{a}{b}.$$

It follows that θ_1 lies in the range $\tan^{-1}(a/b) \leq \theta_1 \leq \pi/2$. Note that θ_1 is *not* the angle between the x-axis and the line joining the origin O to the corresponding point on the curve; for example, when the point is P itself then $\theta_1 = \tan^{-1} a/b$, whilst the line OP makes an angle of $\pi/4$ with the x-axis.

Similarly, referring to that part of

$$\frac{x^2}{b^2} + \frac{y^2}{a^2} = 1$$

that lies on the boundary of the common region as curve C_2, we parameterise it by $x = b \cos \theta_2$ and $y = a \sin \theta_2$ with $0 \leq \theta_2 \leq \tan^{-1}(b/a)$.

On the x-axis, both y and dy are zero and the integrand, $x\, dy - y\, dx$, vanishes. Similarly, the integrand vanishes at all points on the y-axis. Hence,

$$I = \oint_L (x\, dy - y\, dx)$$

$$= \int_{C_2} (x\, dy - y\, dx) + \int_{C_1} (x\, dy - y\, dx)$$

$$= \int_0^{\tan^{-1}(b/a)} [\, ab(\cos \theta_2 \cos \theta_2) - ab \sin \theta_2(-\sin \theta_2) \,]\, d\theta_2$$

$$+ \int_{\tan^{-1}(a/b)}^{\pi/2} [\, ab(\cos \theta_1 \cos \theta_1) - ab \sin \theta_1(-\sin \theta_1) \,]\, d\theta_1$$

$$= ab \tan^{-1} \frac{b}{a} + ab \left(\frac{\pi}{2} - \tan^{-1} \frac{a}{b} \right)$$

$$= 2ab \tan^{-1} \frac{b}{a}.$$

As noted in the question, the area enclosed by L is equal to half of this value, i.e. the total common area in all four quadrants is

$$4 \times \frac{1}{2} \times 2ab \tan^{-1} \frac{b}{a} = 4ab \tan^{-1} \frac{b}{a}.$$

Note that if we let $b \rightarrow a$ then the two ellipses become identical circles and we recover the expected value of πa^2 for their common area.

12.7 Evaluate the line integral

$$I = \oint_C \left[y(4x^2 + y^2)\,dx + x(2x^2 + 3y^2)\,dy \right]$$

around the ellipse $x^2/a^2 + y^2/b^2 = 1$.

As it stands this integral is complicated and, in fact, it is the sum of two integrals. The form of the integrand, containing powers of x and y that can be differentiated easily, makes this problem one to which Green's theorem in a plane might usefully be applied. The theorem states that

$$\oint_C (P\,dx + Q\,dy) = \int\int_R \left(\frac{\partial Q}{\partial x} - \frac{\partial P}{\partial y} \right) dx\,dy,$$

where C is a closed contour enclosing the convex region R.

In the notation used above,

$$P(x, y) = y(4x^2 + y^2) \quad \text{and} \quad Q(x, y) = x(2x^2 + 3y^2).$$

It follows that

$$\frac{\partial P}{\partial y} = 4x^2 + 3y^2 \quad \text{and} \quad \frac{\partial Q}{\partial x} = 6x^2 + 3y^2,$$

leading to

$$\frac{\partial Q}{\partial x} - \frac{\partial P}{\partial y} = 2x^2.$$

This can now be substituted into Green's theorem and the y-integration carried out immediately as the integrand does not contain y. Hence,

$$I = \int\int_R 2x^2\,dx\,dy$$

$$= \int_{-a}^{a} 2x^2\,2b \left(1 - \frac{x^2}{a^2} \right)^{1/2} dx$$

$$= 4b \int_{\pi}^{0} a^2 \cos^2 \phi \, \sin \phi \, (-a \sin \phi \, d\phi), \text{ on setting } x = a \cos \phi,$$

$$= -ba^3 \int_{\pi}^{0} \sin^2(2\phi)\,d\phi = \tfrac{1}{2}\pi ba^3.$$

In the final line we have used the standard result for the integral of the square of a sinusoidal function.

12.9 A single-turn coil C of arbitrary shape is placed in a magnetic field $\mathbf{B}$ and carries a current I. Show that the couple acting upon the coil can be written as

$$\mathbf{M} = I \int_C (\mathbf{B} \cdot \mathbf{r}) \, d\mathbf{r} - I \int_C \mathbf{B}(\mathbf{r} \cdot d\mathbf{r}).$$

For a planar rectangular coil of sides $2a$ and $2b$ placed with its plane vertical and at an angle ϕ to a uniform horizontal field $\mathbf{B}$, show that $\mathbf{M}$ is, as expected, $4abBI \cos \phi \, \mathbf{k}$.

For an arbitrarily shaped coil the total couple acting can only be found by considering that on an infinitesimal element and then integrating this over the whole coil. The force on an element $d\mathbf{r}$ of the coil is $d\mathbf{F} = I \, d\mathbf{r} \times \mathbf{B}$, and the moment of this force about the origin is $d\mathbf{M} = \mathbf{r} \times \mathbf{F}$. Thus the total moment is given by

$$\mathbf{M} = \oint_C \mathbf{r} \times (I \, d\mathbf{r} \times \mathbf{B})$$

$$= I \oint_C (\mathbf{r} \cdot \mathbf{B}) \, d\mathbf{r} - I \oint_C \mathbf{B}(\mathbf{r} \cdot d\mathbf{r}).$$

To obtain this second form we have used the vector identity

$$\mathbf{a} \times (\mathbf{b} \times \mathbf{c}) = (\mathbf{a} \cdot \mathbf{c})\mathbf{b} - (\mathbf{a} \cdot \mathbf{b})\mathbf{c}.$$

To determine the couple acting on the rectangular coil we work in Cartesian coordinates with the z-axis vertical and choose the orientation of axes in the horizontal plane such that the edge of the rectangle of length $2a$ is in the x-direction. Then

$$\mathbf{B} = B \cos \phi \, \mathbf{i} + B \sin \phi \, \mathbf{j}.$$

In the first term in $\mathbf{M}$:

(i) for the horizontal sides

$$\mathbf{r} = x \mathbf{i} \pm b \mathbf{k}, \quad d\mathbf{r} = dx \, \mathbf{i}, \quad \mathbf{r} \cdot \mathbf{B} = x B \cos \phi,$$

$$\int (\mathbf{r} \cdot \mathbf{B}) \, d\mathbf{r} = B \cos \phi \, \mathbf{i} \left(\int_{-a}^{a} x \, dx + \int_{a}^{-a} x \, dx \right) = 0;$$

(ii) for the vertical sides

$$\mathbf{r} = \pm a \mathbf{i} + z \mathbf{k}, \quad d\mathbf{r} = dz \, \mathbf{k}, \quad \mathbf{r} \cdot \mathbf{B} = \pm a B \cos \phi,$$

$$\int (\mathbf{r} \cdot \mathbf{B}) \, d\mathbf{r} = B \cos \phi \, \mathbf{k} \left(\int_{-b}^{b} (+a) \, dz + \int_{b}^{-b} (-a) \, dz \right) = 4abB \cos \phi \, \mathbf{k}.$$

For the second term in $\mathbf{M}$, since the field is uniform it can be taken outside the integral as a (vector) constant. On the horizontal sides the remaining integral is

$$\int \mathbf{r} \cdot d\mathbf{r} = \pm \int_{-a}^{a} x \, dx = 0.$$

Similarly, the contribution from the vertical sides vanishes and the whole of the second term contributes nothing in this particular configuration.

The total moment is thus $4abBI \cos \phi \, \mathbf{k}$, as expected.

12.11 An axially symmetric solid body with its axis AB vertical is immersed in an incompressible fluid of density ρ_0. Use the following method to show that, whatever the shape of the body, for $\rho = \rho(z)$ in cylindrical polars the Archimedean upthrust is, as expected, $\rho_0 g V$, where V is the volume of the body.

Express the vertical component of the resultant force $(-\int p\, d\mathbf{S}$, where p is the pressure) on the body in terms of an integral; note that $p = -\rho_0 g z$ and that for an annular surface element of width dl, $\mathbf{n} \cdot \mathbf{n}_z\, dl = -d\rho$. Integrate by parts and use the fact that $\rho(z_A) = \rho(z_B) = 0$.

We measure z negatively from the water's surface $z = 0$ so that the hydrostatic pressure is $p = -\rho_0 g z$. By symmetry, there is no net horizontal force acting on the body.

The upward force, F, is due to the net vertical component of the hydrostatic pressure acting upon the body's surface:

$$F = -\hat{\mathbf{n}}_z \cdot \int p\, d\mathbf{S}$$

$$= -\hat{\mathbf{n}}_z \cdot \int (-\rho_0 g z)(2\pi \rho\, \hat{\mathbf{n}}\, dl),$$

where $2\pi \rho\, dl$ is the area of the strip of surface lying between z and $z + dz$ and $\hat{\mathbf{n}}$ is the outward unit normal to that surface.

Now, from geometry, $\hat{\mathbf{n}}_z \cdot \hat{\mathbf{n}}$ is equal to minus the sine of the angle between dl and dz and so $\hat{\mathbf{n}}_z \cdot \hat{\mathbf{n}}\, dl$ is equal to $-d\rho$. Thus,

$$F = 2\pi \rho_0 g \int_{z_A}^{z_B} \rho z (-d\rho)$$

$$= -2\pi \rho_0 g \int_{z_A}^{z_B} \left(\rho \frac{\partial \rho}{\partial z} \right) z\, dz$$

$$= -2\pi \rho_0 g \left\{ \left[z \frac{\rho^2}{2} \right]_{z_A}^{z_B} - \int_{z_A}^{z_B} \frac{\rho^2}{2}\, dz \right\}.$$

But $\rho(z_A) = \rho(z_B) = 0$, and so the first contribution vanishes, leaving

$$F = \rho_0 g \int_{z_A}^{z_B} \pi \rho^2\, dz = \rho_0 g V,$$

where V is the volume of the solid. This is the mathematical form of Archimedes' principle. Of course, the result is also valid for a closed body of arbitrary shape, $\rho = \rho(z, \phi)$, but a different method would be needed to prove it.

12.13 A vector field $\mathbf{a}$ is given by $-zxr^{-3}\mathbf{i} - zyr^{-3}\mathbf{j} + (x^2 + y^2)r^{-3}\mathbf{k}$, where $r^2 = x^2 + y^2 + z^2$. Establish that the field is conservative (a) by showing that $\nabla \times \mathbf{a} = 0$ and (b) by constructing its potential function ϕ.

We are told that

$$\mathbf{a} = -\frac{zx}{r^3}\mathbf{i} - \frac{zy}{r^3}\mathbf{j} + \frac{x^2 + y^2}{r^3}\mathbf{k},$$

with $r^2 = x^2 + y^2 + z^2$. We will need to differentiate r^{-3} with respect to x, y and z, using the chain rule, and so note that $\partial r / \partial x = x/r$, etc.

(a) Consider $\nabla \times \mathbf{a}$, term by term:

$$[\nabla \times \mathbf{a}]_x = \frac{\partial}{\partial y} \left(\frac{x^2 + y^2}{r^3} \right) - \frac{\partial}{\partial z} \left(\frac{-zy}{r^3} \right)$$

$$= \frac{-3(x^2 + y^2)y}{r^4 r} + \frac{2y}{r^3} + \frac{y}{r^3} - \frac{3(zy)z}{r^4 r}$$

$$= \frac{3y}{r^5} (-x^2 - y^2 + x^2 + y^2 + z^2 - z^2) = 0;$$

$$[\nabla \times \mathbf{a}]_y = \frac{\partial}{\partial z} \left(\frac{-zx}{r^3} \right) - \frac{\partial}{\partial x} \left(\frac{x^2 + y^2}{r^3} \right)$$

$$= \frac{3(zx)z}{r^4 r} - \frac{x}{r^3} - \frac{2x}{r^3} + \frac{3(x^2 + y^2)x}{r^4 r}$$

$$= \frac{3x}{r^5} (z^2 - x^2 - y^2 - z^2 + x^2 + y^2) = 0;$$

$$[\nabla \times \mathbf{a}]_z = \frac{\partial}{\partial x} \left(\frac{-zy}{r^3} \right) - \frac{\partial}{\partial y} \left(\frac{-zx}{r^3} \right)$$

$$= \frac{3(zy)x}{r^4 r} - \frac{3(zx)y}{r^4 r} = 0.$$

Thus all three components of $\nabla \times \mathbf{a}$ are zero, showing that $\mathbf{a}$ is a conservative field.

(b) To construct its potential function we proceed as follows:

$$\frac{\partial \phi}{\partial x} = \frac{-zx}{(x^2 + y^2 + z^2)^{3/2}} \Rightarrow \phi = \frac{z}{(x^2 + y^2 + z^2)^{1/2}} + f(y, z),$$

$$\frac{\partial \phi}{\partial y} = \frac{-zy}{(x^2 + y^2 + z^2)^{3/2}} = \frac{-zy}{(x^2 + y^2 + z^2)^{3/2}} + \frac{\partial f}{\partial y} \Rightarrow f(y, z) = g(z),$$

$$\frac{\partial \phi}{\partial z} = \frac{x^2 + y^2}{(x^2 + y^2 + z^2)^{3/2}}$$

$$= \frac{1}{(x^2 + y^2 + z^2)^{1/2}} + \frac{-z z}{(x^2 + y^2 + z^2)^{3/2}} + \frac{\partial g}{\partial z}$$

$$\Rightarrow g(z) = c.$$

Thus,

$$\phi(x, y, z) = c + \frac{z}{(x^2 + y^2 + z^2)^{1/2}} = c + \frac{z}{r}.$$

The very fact that we can construct a potential function $\phi = \phi(x, y, z)$ whose derivatives are the components of the vector field shows that the field is conservative.

12.15 A force $\mathbf{F}(\mathbf{r})$ acts on a particle at $\mathbf{r}$. In which of the following cases can $\mathbf{F}$ be represented in terms of a potential? Where it can, find the potential.

(a) $\mathbf{F} = F_0 \left[\mathbf{i} - \mathbf{j} - \dfrac{2(x-y)}{a^2} \mathbf{r} \right] \exp\left(-\dfrac{r^2}{a^2}\right);$

(b) $\mathbf{F} = \dfrac{F_0}{a} \left[z\mathbf{k} + \dfrac{(x^2 + y^2 - a^2)}{a^2} \mathbf{r} \right] \exp\left(-\dfrac{r^2}{a^2}\right);$

(c) $\mathbf{F} = F_0 \left[\mathbf{k} + \dfrac{a(\mathbf{r} \times \mathbf{k})}{r^2} \right].$

(a) We first write the field entirely in terms of the Cartesian unit vectors using $\mathbf{r} = x\mathbf{i} + y\mathbf{j} + z\mathbf{k}$ and then attempt to construct a suitable potential function ϕ:

$$\mathbf{F} = F_0 \left[\mathbf{i} - \mathbf{j} - \frac{2(x-y)}{a^2}\mathbf{r} \right] \exp\left(-\frac{r^2}{a^2}\right)$$

$$= \frac{F_0}{a^2} \left[(a^2 - 2x^2 + 2xy)\mathbf{i} + (-a^2 - 2xy + 2y^2)\mathbf{j} \right.$$

$$\left. + (-2xz + 2yz)\mathbf{k} \right] \exp\left(-\frac{r^2}{a^2}\right).$$

Since the partial derivative of $\exp(-r^2/a^2)$ with respect to any Cartesian coordinate u is $\exp(-r^2/a^2)(-2r/a^2)(u/r)$, the z-component of $\mathbf{F}$ appears to be the most straightforward to tackle first:

$$\frac{\partial \phi}{\partial z} = \frac{F_0}{a^2}(-2xz + 2yz)\exp\left(-\frac{r^2}{a^2}\right)$$

$$\Rightarrow \phi(x, y, z) = F_0(x - y)\exp\left(-\frac{r^2}{a^2}\right) + f(x, y)$$

$$\equiv \phi_1(x, y, z) + f(x, y).$$

Next we examine the derivatives of $\phi = \phi_1 + f$ with respect to x and y to see how closely they generate F_x and F_y:

$$\frac{\partial \phi_1}{\partial x} = F_0 \left[\exp\left(-\frac{r^2}{a^2}\right) + (x-y)\exp\left(-\frac{r^2}{a^2}\right)\left(\frac{-2x}{a^2}\right) \right]$$

$$= \frac{F_0}{a^2}(a^2 - 2x^2 + 2xy)\exp(-r^2/a^2) = F_x \quad \text{(as given),}$$

$$\text{and} \quad \frac{\partial \phi_1}{\partial y} = F_0 \left[-\exp\left(-\frac{r^2}{a^2}\right) + (x-y)\exp\left(-\frac{r^2}{a^2}\right)\left(\frac{-2y}{a^2}\right) \right]$$

$$= \frac{F_0}{a^2}(-a^2 - 2xy + 2y^2)\exp(-r^2/a^2) = F_y \quad \text{(as given).}$$

Thus, to within an arbitrary constant, $\phi_1(x, y, z) = F_0(x - y)\exp\left(-\dfrac{r^2}{a^2}\right)$ is a suitable potential function for the field, without the need for any additional function $f(x, y)$.

(b) We follow the same line of argument as in part (a). First expressing **F** in terms of **i**, **j** and **k**,

$$\mathbf{F} = \frac{F_0}{a}\left[z\,\mathbf{k} + \frac{x^2 + y^2 - a^2}{a^2}\mathbf{r}\right]\exp\left(-\frac{r^2}{a^2}\right)$$

$$= \frac{F_0}{a^3}\left[x(x^2 + y^2 - a^2)\mathbf{i} + y(x^2 + y^2 - a^2)\mathbf{j} + z(x^2 + y^2)\mathbf{k}\right]\exp\left(-\frac{r^2}{a^2}\right),$$

and then constructing a possible potential function ϕ. Again starting with the z-component:

$$\frac{\partial\phi}{\partial z} = \frac{F_0 z}{a^3}(x^2 + y^2)\exp\left(-\frac{r^2}{a^2}\right),$$

$$\Rightarrow \quad \phi(x, y, z) = -\frac{F_0}{2a}(x^2 + y^2)\exp\left(-\frac{r^2}{a^2}\right) + f(x, y)$$

$$\equiv \phi_1(x, y, z) + f(x, y),$$

then,

$$\frac{\partial\phi_1}{\partial x} = -\frac{F_0}{2a}\left[2x - \frac{2x(x^2 + y^2)}{a^2}\right]\exp\left(-\frac{r^2}{a^2}\right) = F_x \quad \text{(as given),}$$

and

$$\frac{\partial\phi_1}{\partial y} = -\frac{F_0}{2a}\left[2y - \frac{2y(x^2 + y^2)}{a^2}\right]\exp\left(-\frac{r^2}{a^2}\right) = F_y \quad \text{(as given).}$$

Thus, $\phi_1(x, y, z) = \dfrac{F_0}{2a}(x^2 + y^2)\exp\left(-\dfrac{r^2}{a^2}\right)$, as it stands, is a suitable potential function for $\mathbf{F}(\mathbf{r})$ and establishes the conservative nature of the field.

(c) Again we express F in Cartesian components:

$$\mathbf{F} = F_0\left[\mathbf{k} + \frac{a(\mathbf{r} \times \mathbf{k})}{r^2}\right] = \frac{ay}{r^2}\mathbf{i} - \frac{ax}{r^2}\mathbf{j} + \mathbf{k}.$$

That the z-component of **F** has no dependence on y whilst its y-component does depend upon z suggests that the x-component of $\nabla \times \mathbf{F}$ may not be zero. To test this out we compute

$$(\nabla \times \mathbf{F})_x = \frac{\partial(1)}{\partial y} - \frac{\partial}{\partial z}\left(\frac{-ax}{r^2}\right) = 0 - \frac{2axz}{r^4} \neq 0$$

and find that it is not. To have even one component of $\nabla \times \mathbf{F}$ non-zero is sufficient to show that **F** is not conservative and that no potential function can be found. There is no point in searching further!

The same conclusion can be reached by considering the implication of $F_z = \mathbf{k}$, namely that any possible potential function has to have the form $\phi(x, y, z) = z + f(x, y)$. However, $\partial\phi/\partial x$ is known to be $-ay/r^2 = -ay/(x^2 + y^2 + z^2)$. This yields a contradiction, as it requires $\partial f(x, y)/\partial x$ to depend on z, which is clearly impossible.

12.17 The vector field $\mathbf{f}$ has components $y\mathbf{i} - x\mathbf{j} + \mathbf{k}$ and γ is a curve given parametrically by

$$\mathbf{r} = (a - c + c\cos\theta)\mathbf{i} + (b + c\sin\theta)\mathbf{j} + c^2\theta\mathbf{k}, \quad 0 \le \theta \le 2\pi.$$

Describe the shape of the path γ and show that the line integral $\int_\gamma \mathbf{f} \cdot d\mathbf{r}$ vanishes. Does this result imply that $\mathbf{f}$ is a conservative field?

As θ increases from 0 to 2π, the x- and y-components of $\mathbf{r}$ vary sinusoidally and in quadrature about fixed values $a - c$ and b. Both variations have amplitude c and both return to their initial values when $\theta = 2\pi$. However, the z-component increases monotonically from 0 to a value of $2\pi c^2$. The curve γ is therefore one loop of a circular spiral of radius c and pitch $2\pi c^2$. Its axis is parallel to the z-axis and passes through the points $(a - c, b, z)$.

The line element $d\mathbf{r}$ has components $(-c\sin\theta\,d\theta, c\cos\theta\,d\theta, c^2\,d\theta)$ and so the line integral of f along γ is given by

$$\int_\gamma \mathbf{f} \cdot d\mathbf{r} = \int_0^{2\pi} \left[y(-c\sin\theta) - x(c\cos\theta) + c^2 \right] d\theta$$

$$= \int_0^{2\pi} \left[-c(b + c\sin\theta)\sin\theta - c(a - c + c\cos\theta)\cos\theta + c^2 \right] d\theta$$

$$= \int_0^{2\pi} \left(-bc\sin\theta - c^2\sin^2\theta - c(a - c)\cos\theta - c^2\cos^2\theta + c^2 \right) d\theta$$

$$= 0 - \pi c^2 - 0 - \pi c^2 + 2\pi c^2 = 0.$$

However, this does not imply that $\mathbf{f}$ is a conservative field since (i) γ is not a closed loop and (ii), even if it were, the line integral has to vanish for *every* loop, not just for a particular one.

Further,

$$\nabla \times \mathbf{f} = (0 - 0,\ 0 - 0,\ -1 - 1) = (0,\ 0,\ -2) \ne \mathbf{0},$$

showing explicitly that $\mathbf{f}$ is not conservative.

12.19 Evaluate the surface integral $\int \mathbf{r} \cdot d\mathbf{S}$, where $\mathbf{r}$ is the position vector, over that part of the surface $z = a^2 - x^2 - y^2$ for which $z \ge 0$, by each of the following methods.

(a) Parameterise the surface as $x = a\sin\theta\cos\phi$, $y = a\sin\theta\sin\phi$, $z = a^2\cos^2\theta$, and show that

$$\mathbf{r} \cdot d\mathbf{S} = a^4(2\sin^3\theta\cos\theta + \cos^3\theta\sin\theta)\,d\theta\,d\phi.$$

(b) Apply the divergence theorem to the volume bounded by the surface and the plane $z = 0$.

(a) With $x = a\sin\theta\cos\phi$, $y = a\sin\theta\sin\phi$, $z = a^2\cos^2\theta$, we first check that this does parameterise the surface appropriately:

$$a^2 - x^2 - y^2 = a^2 - a^2\sin^2\theta(\cos^2\phi + \sin^2\phi) = a^2(1 - \sin^2\theta) = a^2\cos^2\theta = z.$$

We see that it does so for the relevant part of the surface, i.e. that which lies above the plane $z = 0$ with $0 \le \theta \le \pi/2$. It would not do so for the part with $z < 0$ for which $x^2 + y^2$ has to be greater than a^2; this is not catered for by the given parameterisation.

Having carried out this check, we calculate expressions for $d\mathbf{S}$ and hence $\mathbf{r} \cdot d\mathbf{S}$ in terms of θ and ϕ as follows:

$$\mathbf{r} = a \sin\theta \cos\phi \, \mathbf{i} + a \sin\theta \sin\phi \, \mathbf{j} + a^2 \cos^2\theta \, \mathbf{k},$$

and the tangent vectors at the point (θ, ϕ) on the surface are given by

$$\frac{\partial \mathbf{r}}{\partial \theta} = a \cos\theta \cos\phi \, \mathbf{i} + a \cos\theta \sin\phi \, \mathbf{j} - 2a^2 \cos\theta \sin\theta \, \mathbf{k},$$

$$\frac{\partial \mathbf{r}}{\partial \phi} = -a \sin\theta \sin\phi \, \mathbf{i} + a \sin\theta \cos\phi \, \mathbf{j}.$$

The corresponding vector element of surface area is thus

$$d\mathbf{S} = \frac{\partial \mathbf{r}}{\partial \theta} \times \frac{\partial \mathbf{r}}{\partial \phi}$$
$$= 2a^3 \cos\theta \sin^2\theta \cos\phi \, \mathbf{i} + 2a^3 \cos\theta \sin^2\theta \sin\phi \, \mathbf{j} + a^2 \cos\theta \sin\theta \, \mathbf{k},$$

giving $\mathbf{r} \cdot d\mathbf{S}$ as

$$\mathbf{r} \cdot d\mathbf{S} = 2a^4 \cos\theta \sin^3\theta \cos^2\phi + 2a^4 \cos\theta \sin^3\theta \sin^2\phi + a^4 \cos^3\theta \sin\theta$$
$$= 2a^4 \cos\theta \sin^3\theta + a^4 \cos^3\theta \sin\theta.$$

This is to be integrated over the ranges $0 \le \phi < 2\pi$ and $0 \le \theta \le \pi/2$ as follows:

$$\int \mathbf{r} \cdot d\mathbf{S} = a^4 \int_0^{2\pi} d\phi \int_0^{\pi/2} (2 \sin^3\theta \cos\theta + \cos^3\theta \sin\theta) \, d\theta$$

$$= 2\pi a^4 \left(2 \left[\frac{\sin^4\theta}{4} \right]_0^{\pi/2} + \left[\frac{-\cos^4\theta}{4} \right]_0^{\pi/2} \right)$$

$$= 2\pi a^4 \left(\frac{2}{4} + \frac{1}{4} \right) = \frac{3\pi a^4}{2}.$$

(b) The divergence of the vector field $\mathbf{r}$ is 3, a constant, and so the surface integral $\int \mathbf{r} \cdot d\mathbf{S}$ taken over the complete surface Σ (including the part that lies in the plane $z = 0$) is, by the divergence theorem, equal to three times the volume V of the region bounded by Σ. Now,

$$V = \int_0^{a^2} \pi \rho^2 \, dz = \int_0^{a^2} \pi (a^2 - z) \, dz = \pi (a^4 - \tfrac{1}{2}a^4) = \tfrac{1}{2}\pi a^4,$$

and so $\int_\Sigma \mathbf{r} \cdot d\mathbf{S} = 3\pi a^4/2$.

However, on the part of the surface lying in the plane $z = 0$, $\mathbf{r} = x\mathbf{i} + y\mathbf{j} + 0\mathbf{k}$, whilst $d\mathbf{S} = -dS\,\mathbf{k}$. Consequently, the scalar product $\mathbf{r} \cdot d\mathbf{S} = 0$; in words, for any point on this face its position vector is orthogonal to the normal to the face. The surface integral over this face therefore contributes nothing to the total integral and the value obtained is that due to the curved surface alone, in agreement with the result in (a).

12.21 Use the result

$$\int_V \nabla\phi\, dV = \oint_S \phi\, d\mathbf{S},$$

together with an appropriately chosen scalar function ϕ, to prove that the position vector $\bar{\mathbf{r}}$ of the centre of mass of an arbitrarily shaped body of volume V and uniform density can be written

$$\bar{\mathbf{r}} = \frac{1}{V} \oint_S \tfrac{1}{2} r^2\, d\mathbf{S}.$$

The position vector of the centre of mass is defined by

$$\bar{\mathbf{r}} \int_V \rho\, dV = \int_V \mathbf{r}\rho\, dV.$$

Now, we note that $\mathbf{r}$ can be written as $\nabla(\tfrac{1}{2}r^2)$. Thus, cancelling the constant ρ, we have

$$\bar{\mathbf{r}}\, V = \int_V \nabla(\tfrac{1}{2}r^2)\, dV n = \oint_S \tfrac{1}{2} r^2\, d\mathbf{S}$$

$$\Rightarrow \quad \bar{\mathbf{r}} = \frac{1}{V} \oint_S \tfrac{1}{2} r^2\, d\mathbf{S}.$$

This result provides an alternative method of finding the centre of mass $\bar{z}\mathbf{k}$ of the uniform hemisphere $r = a$, $0 \le \theta \le \pi/2$, $0 \le \phi < 2\pi$. The curved surface contributes $3a/4$ to $\bar{z}$ and the plane surface contributes $-3a/8$, giving $\bar{z} = 3a/8$.

12.23 Demonstrate the validity of the divergence theorem:

(a) by calculating the flux of the vector

$$\mathbf{F} = \frac{\alpha\mathbf{r}}{(r^2 + a^2)^{3/2}}$$

through the spherical surface $|\mathbf{r}| = \sqrt{3}a$;
(b) by showing that

$$\nabla \cdot \mathbf{F} = \frac{3\alpha a^2}{(r^2 + a^2)^{5/2}}$$

and evaluating the volume integral of $\nabla \cdot \mathbf{F}$ over the interior of the sphere $|\mathbf{r}| = \sqrt{3}a$. The substitution $r = a\tan\theta$ will prove useful in carrying out the integration.

(a) The field is radial with

$$\mathbf{F} = \frac{\alpha\, \mathbf{r}}{(r^2 + a^2)^{3/2}} = \frac{\alpha\, r}{(r^2 + a^2)^{3/2}}\,\hat{\mathbf{e}}_r.$$

The total flux is therefore given by

$$\Phi = \left.\frac{4\pi r^2\,\alpha\, r}{(r^2 + a^2)^{3/2}}\right|_{r=\sqrt{3}a} = \frac{4\pi a^3\,\alpha\, 3\sqrt{3}}{8a^3} = \frac{3\sqrt{3}\pi\alpha}{2}.$$

(b) From the divergence theorem, the total flux over the surface of the sphere is equal to the volume integral of its divergence within the sphere. The divergence is given by

$$\nabla \cdot \mathbf{F} = \frac{1}{r^2}\frac{\partial}{\partial r}(r^2 F_r) = \frac{1}{r^2}\frac{\partial}{\partial r}\left(\frac{r^2\, \alpha r}{(r^2 + a^2)^{3/2}}\right)$$

$$= \frac{1}{r^2}\left[\frac{3\alpha r^2}{(r^2 + a^2)^{3/2}} - \frac{3\alpha r^4}{(r^2 + a^2)^{5/2}}\right]$$

$$= \frac{3\alpha a^2}{(r^2 + a^2)^{5/2}}.$$

Since the divergence varies with the radial distance r, in order to find the volume integral of the divergence it is necessary to integrate over r from 0 to $\sqrt{3}a$:

$$\int_V \nabla \cdot \mathbf{F}\, dV = \int_0^{\sqrt{3}a} \frac{3\alpha a^2}{(r^2 + a^2)^{5/2}} 4\pi r^2\, dr.$$

Now setting $r = a \tan\theta$, for $0 \le \theta \le \pi/3$, we have

$$\int_V \nabla \cdot \mathbf{F}\, dV = 12\pi\alpha a^2 \int_0^{\pi/3} \frac{a^2 \tan^2\theta\, a \sec^2\theta}{a^5 \sec^5\theta}\, d\theta$$

$$= 12\pi\alpha \int_0^{\pi/3} \sin^2\theta \cos\theta\, d\theta$$

$$= 12\pi\alpha \left[\frac{\sin^3\theta}{3}\right]_0^{\pi/3} = 12\pi\alpha\frac{\sqrt{3}}{8} = \frac{3\sqrt{3}\pi\alpha}{2}, \quad \text{as in (a).}$$

The equality of the results in parts (a) and (b) is in accordance with the divergence theorem.

12.25 In a uniform conducting medium with unit relative permittivity, charge density ρ, current density $\mathbf{J}$, electric field $\mathbf{E}$ and magnetic field $\mathbf{B}$, Maxwell's electromagnetic equations take the form (with $\mu_0\epsilon_0 = c^{-2}$)

(i) $\nabla \cdot \mathbf{B} = 0$,   (ii) $\nabla \cdot \mathbf{E} = \rho/\epsilon_0$,

(iii) $\nabla \times \mathbf{E} + \dot{\mathbf{B}} = \mathbf{0}$,   (iv) $\nabla \times \mathbf{B} - (\dot{\mathbf{E}}/c^2) = \mu_0\mathbf{J}$.

The density of stored energy in the medium is given by $\frac{1}{2}(\epsilon_0 E^2 + \mu_0^{-1} B^2)$. Show that the rate of change of the total stored energy in a volume V is equal to

$$-\int_V \mathbf{J} \cdot \mathbf{E}\, dV - \frac{1}{\mu_0} \oint_S (\mathbf{E} \times \mathbf{B}) \cdot d\mathbf{S},$$

where S is the surface bounding V.

[The first integral gives the ohmic heating loss, whilst the second gives the electromagnetic energy flux out of the bounding surface. The vector $\mu_0^{-1}(\mathbf{E} \times \mathbf{B})$ is known as the Poynting vector.]

The total stored energy is equal to the volume integral of the energy density. Let R be its rate of change. Then, differentiating under the integral sign, we have

$$R = \frac{d}{dt} \int_V \left(\frac{\epsilon_0}{2} E^2 + \frac{1}{2\mu_0} B^2 \right) dV$$

$$= \int_V \left(\epsilon_0 \mathbf{E} \cdot \dot{\mathbf{E}} + \frac{1}{\mu_0} \mathbf{B} \cdot \dot{\mathbf{B}} \right) dV.$$

Now using (iv) and (iii), we have

$$R = \int_V \left[\epsilon_0 \mathbf{E} \cdot (-\mu_0 c^2 \mathbf{J} + c^2 \nabla \times \mathbf{B}) - \frac{1}{\mu_0} \mathbf{B} \cdot (\nabla \times \mathbf{E}) \right] dV$$

$$= - \int_V \mathbf{E} \cdot \mathbf{J} \, dV + \int_V \left[\epsilon_0 c^2 \, \mathbf{E} \cdot (\nabla \times \mathbf{B}) - \frac{1}{\mu_0} \mathbf{B} \cdot (\nabla \times \mathbf{E}) \right] dV$$

$$= - \int_V \mathbf{E} \cdot \mathbf{J} \, dV - \frac{1}{\mu_0} \int_V \nabla \cdot (\mathbf{E} \times \mathbf{B}) \, dV$$

$$= - \int_V \mathbf{E} \cdot \mathbf{J} \, dV - \frac{1}{\mu_0} \oint_S (\mathbf{E} \times \mathbf{B}) \cdot d\mathbf{S}.$$

We used the vector identity

$$\nabla \cdot (\mathbf{a} \times \mathbf{b}) = \mathbf{b} \cdot (\nabla \times \mathbf{a}) - \mathbf{a} \cdot (\nabla \times \mathbf{b}).$$

to obtain the penultimate line and the divergence theorem to obtain the final one.

12.27 The vector field $\mathbf{F}$ is given by
$$\mathbf{F} = (3x^2 yz + y^3 z + xe^{-x})\mathbf{i} + (3xy^2 z + x^3 z + ye^x)\mathbf{j} + (x^3 y + y^3 x + xy^2 z^2)\mathbf{k}.$$

Calculate (a) directly and (b) by using Stokes' theorem the value of the line integral $\int_L \mathbf{F} \cdot d\mathbf{r}$, where L is the (three-dimensional) closed contour $OABCDEO$ defined by the successive vertices $(0, 0, 0)$, $(1, 0, 0)$, $(1, 0, 1)$, $(1, 1, 1)$, $(1, 1, 0)$, $(0, 1, 0)$, $(0, 0, 0)$.

(a) This calculation is a piece-wise evaluation of the line integral, made up of a series of scalar products of the length of a straight piece of the contour and the component of $\mathbf{F}$ parallel to it (integrated if that component varies along the particular straight section).

On OA, $y = z = 0$ and $F_x = xe^{-x}$;

$$I_1 = \int_0^1 xe^{-x} \, dx = \left[-xe^{-x} \right]_0^1 + \int_0^1 e^{-x} \, dx = 1 - 2e^{-1}.$$

On AB, $x = 1$ and $y = 0$ and $F_z = 0$; the integral I_2 is zero.

On BC, $x = 1$ and $z = 1$ and $F_y = 3y^2 + 1 + ey$;

$$I_3 = \int_0^1 (3y^2 + 1 + ey) \, dy = 1 + 1 + \tfrac{1}{2}e.$$

On CD, $x = 1$ and $y = 1$ and $F_z = 1 + 1 + z^2$;

$$I_4 = \int_1^0 (1 + 1 + z^2)\,dz = -1 - 1 - \tfrac{1}{3}.$$

On DE, $y = 1$ and $z = 0$ and $F_x = xe^{-x}$;

$$I_5 = \int_1^0 xe^{-x}\,dx = -1 + 2e^{-1}.$$

On EO, $x = z = 0$ and $F_y = ye^0$;

$$I_6 = \int_1^0 ye^0\,dy = -\tfrac{1}{2}.$$

Adding up these six contributions shows that the complete line integral has the value $\dfrac{e}{2} - \dfrac{5}{6}$.

(b) As a simple sketch shows, the given contour is three-dimensional. However, it is equivalent to two plane square contours, one $OADEO$ (denoted by S_1) lying in the plane $z = 0$ and the other $ABCDA$ (S_2) lying in the plane $x = 1$; the latter is traversed in the negative sense. The common segment AD does not form part of the original contour but, as it is traversed in opposite senses in the two constituent contours, it (correctly) contributes nothing to the line integral.

To use Stokes' theorem we first need to calculate

$$(\nabla \times \mathbf{F})_x = x^3 + 3y^2x + 2yxz^2 - 3xy^2 - x^3 = 2yxz^2,$$
$$(\nabla \times \mathbf{F})_y = 3x^2y + y^3 - 3x^2y - y^3 - y^2z^2 = -y^2z^2,$$
$$(\nabla \times \mathbf{F})_z = 3y^2z + 3x^2z + ye^x - 3x^2z - 3y^2z = ye^x.$$

Now, S_1 has its normal in the positive z-direction, and so only the z-component of $\nabla \times \mathbf{F}$ is needed in the first surface integral of Stokes' theorem. Likewise, only the x-component of $\nabla \times \mathbf{F}$ is needed in the second integral, but its value must be subtracted because of the sense in which its contour is traversed:

$$\int_{OABCDEO} (\nabla \times \mathbf{F}) \cdot d\mathbf{r} = \int_{S_1} (\nabla \times \mathbf{F})_z\,dx\,dy - \int_{S_2} (\nabla \times \mathbf{F})_x\,dy\,dz$$

$$= \int_0^1 \int_0^1 ye^x\,dx\,dy - \int_0^1 \int_0^1 2y \times 1 \times z^2\,dy\,dz$$

$$= \frac{1}{2}(e - 1) - 2\frac{1}{2}\frac{1}{3} = \frac{e}{2} - \frac{5}{6}.$$

As they must, the two methods give the same value.

13 Laplace transforms

13.1 Find the Laplace transforms of $t^{-1/2}$ and $t^{1/2}$, by setting $x^2 = ts$ in the result

$$\int_0^\infty \exp(-x^2)\, dx = \tfrac{1}{2}\sqrt{\pi}.$$

Setting $x^2 = st$, and hence $2x\, dx = s\, dt$ and $dx = s\, dt/(2\sqrt{st})$, we obtain

$$\int_0^\infty e^{-st}\, \frac{\sqrt{s}}{2}\, t^{-1/2}\, dt = \frac{\sqrt{\pi}}{2},$$

$$\Rightarrow \quad \mathcal{L}\left[t^{-1/2}\right] \equiv \int_0^\infty t^{-1/2}\, e^{-st}\, dt = \sqrt{\frac{\pi}{s}}.$$

Integrating the LHS of this result by parts yields

$$\left[e^{-st}\, 2t^{1/2}\right]_0^\infty - \int_0^\infty (-s)\, e^{-st}\, 2t^{1/2}\, dt = \sqrt{\frac{\pi}{s}}.$$

The first term vanishes at both limits, whilst the second is a multiple of the required Laplace transform of $t^{1/2}$. Hence,

$$\mathcal{L}\left[t^{1/2}\right] \equiv \int_0^\infty e^{-st}\, t^{1/2}\, dt = \frac{1}{2s}\sqrt{\frac{\pi}{s}}.$$

13.3 Use the properties of Laplace transforms to prove the following without evaluating any Laplace integrals explicitly:

(a) $\mathcal{L}\left[t^{5/2}\right] = \frac{15}{8}\sqrt{\pi}\, s^{-7/2}$;

(b) $\mathcal{L}\left[(\sinh at)/t\right] = \frac{1}{2}\ln\left[(s+a)/(s-a)\right], \qquad s > |a|$;

(c) $\mathcal{L}\left[\sinh at \cos bt\right] = a(s^2 - a^2 + b^2)[(s-a)^2 + b^2]^{-1}[(s+a)^2 + b^2]^{-1}$.

(a) We use the general result for Laplace transforms that

$$\mathcal{L}\left[t^n f(t)\right] = (-1)^n \frac{d^n \bar{f}(s)}{ds^n}, \qquad \text{for } n = 1, 2, 3, \ldots$$

If we take $n = 2$, then $f(t)$ becomes $t^{1/2}$, for which we found the Laplace transform in Problem 13.1:

$$\mathcal{L}[t^{5/2}] = \mathcal{L}[t^2 t^{1/2}] = (-1)^2 \frac{d^2}{ds^2}\left(\frac{\sqrt{\pi}\, s^{-3/2}}{2}\right)$$

$$= \frac{\sqrt{\pi}}{2}\left(-\frac{3}{2}\right)\left(-\frac{5}{2}\right) s^{-7/2} = \frac{15\sqrt{\pi}}{8} s^{-7/2}.$$

(b) Here we apply a second general result for Laplace transforms which states that

$$\mathcal{L}\left[\frac{f(t)}{t}\right] = \int_s^\infty \bar{f}(u)\, du,$$

provided $\lim_{t\to 0}[\, f(t)/t\,]$ exists, which it does in this case.

$$\mathcal{L}\left[\frac{\sinh(at)}{t}\right] = \int_s^\infty \frac{a}{u^2 - a^2}\, du, \quad u > |a|,$$

$$= \frac{1}{2}\int_s^\infty \left(\frac{1}{u-a} - \frac{1}{u+a}\right) du$$

$$= \frac{1}{2}\ln\left(\frac{s+a}{s-a}\right), \quad s > |a|.$$

(c) The translation property of Laplace transforms can be used here to deal with the $\sinh(at)$ factor, as it can be expressed in terms of exponential functions:

$$\mathcal{L}[\sinh(at)\cos(bt)] = \mathcal{L}\left[\tfrac{1}{2}e^{at}\cos(bt)\right] - \mathcal{L}\left[\tfrac{1}{2}e^{-at}\cos(bt)\right]$$

$$= \frac{1}{2}\frac{s-a}{(s-a)^2 + b^2} - \frac{1}{2}\frac{s+a}{(s+a)^2 + b^2}$$

$$= \frac{1}{2}\frac{(s^2 - a^2)2a + 2ab^2}{[(s-a)^2 + b^2][(s+a)^2 + b^2]}$$

$$= \frac{a(s^2 - a^2 + b^2)}{[(s-a)^2 + b^2][(s+a)^2 + b^2]}.$$

The result is valid for $s > |a|$.

13.5 This problem is concerned with the limiting behaviour of Laplace transforms.

(a) If $f(t) = A + g(t)$, where A is a constant and the indefinite integral of $g(t)$ is bounded as its upper limit tends to ∞, show that

$$\lim_{s\to 0} s\bar{f}(s) = A.$$

(b) For $t > 0$, the function $y(t)$ obeys the differential equation

$$\frac{d^2 y}{dt^2} + a\frac{dy}{dt} + by = c\cos^2 \omega t,$$

where a, b and c are positive constants. Find $\bar{y}(s)$ and show that $s\bar{y}(s) \to c/2b$ as $s \to 0$. Interpret the result in the t-domain.

(a) From the definition,

$$\bar{f}(s) = \int_0^\infty [A + g(t)] \, e^{-st} \, dt$$

$$= \left[\frac{A e^{-st}}{-s} \right]_0^\infty + \lim_{T \to \infty} \int_0^T g(t) \, e^{-st} \, dt,$$

$$s \bar{f}(s) = A + s \lim_{T \to \infty} \int_0^T g(t) \, e^{-st} \, dt.$$

Now, for $s \geq 0$,

$$\left| \lim_{T \to \infty} \int_0^T g(t) \, e^{-st} \, dt \right| \leq \left| \lim_{T \to \infty} \int_0^T g(t) \, dt \right| < B, \text{ say.}$$

Thus, taking the limit $s \to 0$,

$$\lim_{s \to 0} s \bar{f}(s) = A \pm \lim_{s \to 0} s B = A.$$

(b) We will need

$$\mathcal{L}\left[\cos^2 \omega t \right] = \mathcal{L}\left[\tfrac{1}{2} \cos 2\omega + \tfrac{1}{2} \right] = \frac{s}{2(s^2 + 4\omega^2)} + \frac{1}{2s}.$$

Taking the transform of the differential equation yields

$$-y'(0) - s y(0) + s^2 \bar{y} + a[-y(0) + s\bar{y}] + b\bar{y} = c \left[\frac{s}{2(s^2 + 4\omega^2)} + \frac{1}{2s} \right].$$

This can be rearranged as

$$s\bar{y} = \frac{c \left[\dfrac{s^2}{2(s^2 + 4\omega^2)} + \dfrac{1}{2} \right] + s y'(0) + a s y(0) + s^2 y(0)}{s^2 + as + b}.$$

In the limit $s \to 0$, this tends to $(c/2)/b = c/(2b)$, a value independent of that of a and the initial values of y and y'.

The $s = 0$ component of the transform corresponds to long-term values, when a steady state has been reached and rates of change are negligible. With the first two terms of the differential equation ignored, it reduces to $by = c \cos^2 \omega t$, and, as the average value of $\cos^2 \omega t$ is $\tfrac{1}{2}$, the solution is the more or less steady value of $y = \tfrac{1}{2} c/b$.

13.7 The function $f_a(x)$ is defined as unity for $0 < x < a$ and zero otherwise. Find its Laplace transform $\bar{f}_a(s)$ and deduce that the transform of $x f_a(x)$ is

$$\frac{1}{s^2} \left[1 - (1 + as) e^{-sa} \right].$$

Write $f_a(x)$ in terms of Heaviside functions and hence obtain an explicit expression for

$$g_a(x) = \int_0^x f_a(y)f_a(x-y)\,dy.$$

Use the expression to write $\bar{g}_a(s)$ in terms of the functions $\bar{f}_a(s)$ and $\bar{f}_{2a}(s)$, and their derivatives, and hence show that $\bar{g}_a(s)$ is equal to the square of $\bar{f}_a(s)$, in accordance with the convolution theorem.

From their definitions,

$$\bar{f}_a(s) = \int_0^a 1\, e^{-sx}\, dx = \frac{1}{s}(1 - e^{-sa}),$$

$$\int_0^a x\, f_a(x)\, e^{-sx}\, dx = -\frac{d\bar{f}_a}{ds} = \frac{1}{s^2}(1 - e^{-sa}) - \frac{a}{s}e^{-sa}$$

$$= \frac{1}{s^2}\left[1 - (1 + as)e^{-sa}\right]. \qquad (*)$$

In terms of Heaviside functions,

$$f(x) = H(x) - H(x - a),$$

and so the expression for $g_a(x) = \int_0^x f_a(y)f_a(x-y)\,dy$ is

$$\int_{-\infty}^{\infty} [\,H(y) - H(y-a)\,][\,H(x-y) - H(x-y-a)\,]\,dy.$$

This can be expanded as the sum of four integrals, each of which contains the common factors $H(y)$ and $H(x - y)$, implying that, in all cases, unless x is positive and greater than y, the integral has zero value. The other factors in the four integrands are generated analogously to the terms of the expansion $(a - b)(c - d) = ac - ad - bc + bd$:

$$\int_{-\infty}^{\infty} H(y)H(x-y)\,dy$$

$$-\int_{-\infty}^{\infty} H(y)H(x-y-a)\,dy$$

$$-\int_{-\infty}^{\infty} H(y-a)H(x-y)\,dy$$

$$+\int_{-\infty}^{\infty} H(y-a)H(x-y-a)\,dy.$$

In all four integrals the integrand is either 0 or 1 and the value of each integral is equal to the length of the y-interval in which the integrand is non-zero.

- The first integral requires $0 < y < x$ and therefore has value x for $x > 0$.
- The second integral requires $0 < y < x - a$ and therefore has value $x - a$ for $x > a$ and 0 for $x < a$.

- The third integral requires $a < y < x$ and therefore has value $x - a$ for $x > a$ and 0 for $x < a$.
- The final integral requires $a < y < x - a$ and therefore has value $x - 2a$ for $x > 2a$ and 0 for $x < 2a$.

Collecting these together:

$$
\begin{aligned}
x < 0 \qquad & g_a(x) = 0 - 0 - 0 + 0 = 0, \\
0 < x < a \qquad & g_a(x) = x - 0 - 0 + 0 = x, \\
a < x < 2a \qquad & g_a(x) = x - (x - a) - (x - a) + 0 = 2a - x, \\
2a < x \qquad & g_a(x) = x - (x - a) - (x - a) + (x - 2a) = 0.
\end{aligned}
$$

Consequently, the transform of $g_a(x)$ is given by

$$
\begin{aligned}
\bar{g}_a(s) &= \int_0^a x e^{-sx}\, dx + \int_a^{2a} (2a - x) e^{-sx}\, dx \\[2mm]
&= -\int_0^{2a} x e^{-sx}\, dx + 2\int_0^a x e^{-sx}\, dx + 2a \int_a^{2a} e^{-sx}\, dx \\[2mm]
&= -\frac{1}{s^2}\left[1 - (1 + 2as)e^{-2sa}\right] + \frac{2}{s^2}\left[1 - (1 + as)e^{-sa}\right] + \frac{2a}{s}(e^{-sa} - e^{-2sa}) \\[2mm]
&= \frac{1}{s^2}(1 - 2e^{-sa} + e^{-2sa}) \\[2mm]
&= \frac{1}{s^2}(1 - e^{-as})^2 = [\,\bar{f}_a(s)\,]^2,
\end{aligned}
$$

which is as expected. In order to adjust the integral limits in the second line, we both added and subtracted

$$
\int_0^a (-x)e^{-sx}\, dx.
$$

In the third line we used the result (∗) twice, once as it stands and once with a replaced by $2a$.

14

Ordinary differential equations

14.1 A radioactive isotope decays in such a way that the number of atoms present at a given time, $N(t)$, obeys the equation

$$\frac{dN}{dt} = -\lambda N.$$

If there are initially N_0 atoms present, find $N(t)$ at later times.

This is a straightforward separable equation with a well-known solution:

$$\frac{dN}{dt} = -\lambda N.$$

Separating the variables,
$$\frac{dN}{N} = -\lambda \, dt.$$

Integrating,
$$\ln N(t) - \ln N(0) = -\lambda(t - 0).$$

Thus, since $N(0) = N_0$, we have that, at a later time,

$$N(t) = N_0 e^{-\lambda t}.$$

14.3 Show that the following equations either are exact or can be made exact, and solve them:

(a) $y(2x^2y^2 + 1)y' + x(y^4 + 1) = 0$;
(b) $2xy' + 3x + y = 0$;
(c) $(\cos^2 x + y \sin 2x)y' + y^2 = 0$.

In general, given an equation expressed in the form $A \, dx + B \, dy = 0$, we consider the function

$$h(x, y) = \frac{1}{B}\left[\frac{\partial A}{\partial y} - \frac{\partial B}{\partial x}\right].$$

If this expression is zero, then the equation is exact and can be integrated as it stands to give a solution of the form $f(x, y) = c$. Even if $g(x, y)$ is non-zero, if it is a function of x alone then

$$\mu(x) = \exp\left\{\int g(x) \, dx\right\}$$

provides an integrating factor (IF) that will make the equation exact. Similar considerations apply if $g(x, y)$ is a function of y alone. If g does actually depend on both x and y, then, in general, no further progress can be made using this method.

(a) Following the above procedure, we consider

$$h(x, y) = \frac{1}{2x^2y^3 + y} \left[\frac{\partial}{\partial y} (xy^4 + x) - \frac{\partial}{\partial x} (2x^2y^3 + y) \right] = \frac{4xy^3 - 4xy^3}{2x^2y^3 + y} = 0.$$

It follows that the equation is exact and can be integrated as it stands:

$$c = f(x, y) = \int (2x^2y^3 + y)\,dy + g(x)$$

$$= \frac{1}{2}x^2y^4 + \frac{1}{2}y^2 + g(x), \text{ where}$$

$$xy^4 + x = \frac{\partial f}{\partial x} = xy^4 + 0 + g'(x), \quad \Rightarrow \quad g(x) = \frac{1}{2}x^2 + k,$$

$$\Rightarrow \quad c = f(x, y) = \frac{1}{2}(x^2y^4 + y^2 + x^2).$$

The common factor of $\frac{1}{2}$ on the RHS can, of course, be absorbed into the constant on the LHS and has no particular significance.

(b) Again following the procedure, we consider

$$h(x, y) = \frac{1}{2x} \left[\frac{\partial}{\partial y} (3x + y) - \frac{\partial}{\partial x} (2x) \right] = -\frac{1}{2x}.$$

This is non-zero and implies that the equation is not exact. However, it is a function of x alone and so there is an IF given by

$$\mu(x) = \exp \left\{ \int -\frac{1}{2x}\,dx \right\} = \exp(-\frac{1}{2}\ln x) = \frac{1}{x^{1/2}}.$$

The exact equation is thus

$$2x^{1/2}\,dy + (3x^{1/2} + yx^{-1/2})\,dx = 0,$$

and this can now be integrated:

$$c = f(x, y) = \int 2x^{1/2}\,dy + g(x)$$

$$= 2x^{1/2}y + g(x), \text{ where}$$

$$3x^{1/2} + yx^{-1/2} = \frac{\partial f}{\partial x} = x^{-1/2}y + g'(x), \quad \Rightarrow \quad g(x) = 2x^{3/2} + k,$$

$$\Rightarrow \quad c = f(x, y) = 2(x^{1/2}y + x^{3/2}).$$

Again, the overall numerical multiplicative factor on the RHS has no particular significance.

(c) Following the same general procedure,

$$h(x, y) = \frac{1}{\cos^2 x + y \sin 2x} \left[\frac{\partial}{\partial y} (y^2) - \frac{\partial}{\partial x} (\cos^2 x + y \sin 2x) \right]$$

$$= \frac{1}{\cos^2 x + y \sin 2x} (2y + \sin 2x - 2y \cos 2x)$$

$$= \frac{4y \sin^2 x + 2 \sin x \cos x}{\cos^2 x + y \sin 2x}$$

$$= \frac{2 \sin x (2y \sin x + \cos x)}{\cos x (\cos x + 2y \sin x)} = 2 \tan x.$$

This is non-zero and implies that the equation is not exact. However, it is a function of x alone and so there is an IF given by

$$\mu(x) = \exp \left\{ \int 2 \tan x \, dx \right\} = \exp(-2 \ln \cos x) = \frac{1}{\cos^2 x}.$$

The exact equation is thus

$$(1 + 2y \tan x) \, dy + y^2 \sec^2 x \, dx = 0,$$

and this can now be integrated:

$$c = f(x, y) = \int (1 + 2y \tan x) \, dy + g(x)$$

$$= y + y^2 \tan x + g(x), \quad \text{where}$$

$$y^2 \sec^2 x = \frac{\partial f}{\partial x} = 0 + y^2 \sec^2 x + g'(x), \quad \Rightarrow \quad g(x) = k,$$

$$\Rightarrow \quad c = f(x, y) = y + y^2 \tan x.$$

14.5 By finding suitable IFs, solve the following equations:

(a) $(1 - x^2)y' + 2xy = (1 - x^2)^{3/2}$;
(b) $y' - y \cot x + \csc x = 0$;
(c) $(x + y^3)y' = y$ (treat y as the independent variable).

(a) In standard form this is

$$y' + \frac{2xy}{1 - x^2} = (1 - x^2)^{1/2}.$$

The IF for this standard form is

$$\mu(x) = \exp \left\{ \int \frac{2x}{1 - x^2} \, dx \right\} = \exp[-\ln(1 - x^2)] = \frac{1}{1 - x^2},$$

i.e. $(1 - x^2)^{-2}$ for the original form. Applying it gives

$$\frac{y'}{1 - x^2} + \frac{2xy}{(1 - x^2)^2} = \frac{1}{(1 - x^2)^{1/2}},$$

$$\frac{d}{dx}\left(\frac{y}{1 - x^2}\right) = \frac{1}{(1 - x^2)^{1/2}},$$

$$\frac{y}{1 - x^2} = \sin^{-1} x + k,$$

$$\Rightarrow \quad y = (1 - x^2)(\sin^{-1} x + k).$$

(b) In standard form this is

$$y' - \frac{y \cos x}{\sin x} = -\frac{1}{\sin x}.$$

The IF for this standard form is given by

$$\mu(x) = \exp\left\{-\int \frac{\cos x}{\sin x}\, dx\right\} = \exp[-\ln(\sin x)] = \frac{1}{\sin x}.$$

Applying it gives

$$\frac{y'}{\sin x} - \frac{y \cos x}{\sin^2 x} = -\frac{1}{\sin^2 x},$$

$$\frac{d}{dx}\left(\frac{y}{\sin x}\right) = -\mathrm{cosec}^2 x,$$

$$\frac{y}{\sin x} = \cot x + k,$$

$$\Rightarrow \quad y = \cos x + k \sin x.$$

(c) Rearranging this to make y the independent variable,

$$\frac{dx}{dy} - \frac{x}{y} = y^2.$$

By inspection (or by the standard method) the IF is y^{-1}, yielding

$$\frac{1}{y}\frac{dx}{dy} - \frac{x}{y^2} = \frac{y^2}{y}$$

$$\frac{d}{dy}\left(\frac{x}{y}\right) = y,$$

$$\frac{x}{y} = \frac{1}{2}y^2 + k,$$

$$\Rightarrow \quad x = \tfrac{1}{2}y^3 + ky.$$

14.7 Find, in the form of an integral, the solution of the equation

$$\alpha \frac{dy}{dt} + y = f(t)$$

for a general function $f(t)$. Find the specific solutions for

(a) $f(t) = H(t)$,
(b) $f(t) = \delta(t)$,
(c) $f(t) = \beta^{-1} e^{-t/\beta} H(t)$ with $\beta < \alpha$.

For case (c), what happens if $\beta \to 0$?

The IF needed for the standard form is $\exp[\int \alpha^{-1} dt]$, i.e. $e^{t/\alpha}$. The equation then reads

$$e^{t/\alpha} \frac{dy}{dt} + \frac{y\, e^{t/\alpha}}{\alpha} = \frac{f(t)\, e^{t/\alpha}}{\alpha},$$

$$\frac{d}{dt}\left(y\, e^{t/\alpha} \right) = \frac{f(t)\, e^{t/\alpha}}{\alpha},$$

$$y(t) = e^{-t/\alpha} \int^{t} \frac{f(t')\, e^{t'/\alpha}}{\alpha}\, dt'.$$

We now apply this general result to the three specific cases.

(a) $f(t) = H(t)$, the Heaviside function. This is zero for $t < 0$ and so we can take the integral as running from 0 to t. The value of $H(t)$ for $t > 0$ is unity. Hence,

$$y(t) = e^{-t/\alpha} \int_{0}^{t} \frac{e^{t'/\alpha}}{\alpha}\, dt' = e^{-t/\alpha}[e^{t/\alpha} - 1] = 1 - e^{-t/\alpha}.$$

(b) With $f(t) = \delta(t)$, the integration will be trivial:

$$y(t) = e^{-t/\alpha} \int^{t} \frac{\delta(t')\, e^{t'/\alpha}}{\alpha}\, dt' = e^{-t/\alpha} \times \frac{1}{\alpha} = \frac{e^{-t/\alpha}}{\alpha}.$$

(c) For $f(t) = \beta^{-1} e^{-t/\beta} H(t)$, with $\beta < \alpha$, we have

$$y(t) = e^{-t/\alpha} \int_{0}^{t} \frac{e^{t'/\alpha}\, e^{-t'/\beta}}{\alpha\beta}\, dt'$$

$$= e^{-t/\alpha} \left[\frac{e^{(\alpha^{-1} - \beta^{-1})t'}}{\alpha\beta(\alpha^{-1} - \beta^{-1})} \right]_{0}^{t}$$

$$= \frac{e^{-t/\beta}}{\beta - \alpha} - \frac{e^{-t/\alpha}}{\beta - \alpha}$$

$$= \frac{e^{-t/\alpha} - e^{-t/\beta}}{\alpha - \beta}.$$

As $\beta \to 0$, $f(t)$ becomes very strongly peaked near $t = 0$, but with the area under the peak remaining constant at unity. In the limit, the input $f(t)$ becomes a δ-function, the same as that in case (b). It can also be seen that in the same limit the solution $y(t)$ for case (c) tends to that for case (b), as is to be expected.

14.9 A two-dimensional coordinate system that is useful for orbit problems is the tangential–polar coordinate system. In this system a curve is defined by r, the distance from a fixed point O to a general point P of the curve, and p, the perpendicular distance from O to the tangent to the curve at P. It can be shown that the instantaneous radius of curvature of the curve is given by $\rho = r\,dr/dp$.

Using tangential–polar coordinates, consider a particle of mass m moving under the influence of a force f directed towards the origin O. By resolving forces along the instantaneous tangent and normal, prove that

$$f = -mv\frac{dv}{dr} \quad \text{and} \quad mv^2 = fp\frac{dr}{dp}.$$

Show further that $h = mpv$ is a constant of the motion and that the law of force can be deduced from

$$f = \frac{h^2}{mp^3}\frac{dp}{dr}.$$

Denote by ϕ the angle between the radius vector and the tangent to the orbit at any instant. Then, firstly, we note that $\cos\phi = dr/ds$, where s is the distance moved along the orbit curve and, secondly, that $p = r\sin\phi$.

Now we equate the tangential component of the central force $-f\cos\phi$ to the rate of change of the tangential momentum:

$$-f\frac{dr}{ds} = -f\cos\phi = m\frac{dv}{dt} = m\frac{dv}{ds}\frac{ds}{dt} = mv\frac{dv}{ds}.$$

Hence,

$$f = -mv\frac{dv}{ds}\frac{ds}{dr} = -mv\frac{dv}{dr}.$$

This is the first of the results.

Equating the normal component of the central force to that needed to keep the particle moving in an orbit with instantaneous radius of curvature $\rho = r\,dr/dp$ gives

$$\frac{mv^2}{\rho} = f\sin\phi = f\frac{p}{r} \quad \Rightarrow \quad mv^2 = f\frac{p}{r}r\frac{dr}{dp} = fp\frac{dr}{dp}.$$

Eliminating f from the two equations yields

$$mv^2 = -mvp\frac{dv}{dp} \quad \Rightarrow \quad mv + mp\frac{dv}{dp} = 0$$

$$\Rightarrow \quad h \equiv mpv \text{ is a constant of the motion.}$$

It follows that

$$f = \frac{mv^2}{p}\frac{dp}{dr} = \frac{h^2}{mp^3}\frac{dp}{dr},$$

from which the law of force can be deduced once p is given as a function of r.

14.11 Solve

$$(y - x)\frac{dy}{dx} + 2x + 3y = 0.$$

We first test whether the equation is exact, or can be made so with the help of an IF. To do this, we write the equation as

$$(y - x)\,dy + (2x + 3y)\,dx = 0$$

and consider

$$h_x(x, y) = \frac{1}{y - x}\left[\frac{\partial}{\partial y}(2x + 3y) - \frac{\partial}{\partial x}(y - x)\right] = \frac{4}{y - x}.$$

This is not a function of x alone. Equally

$$h_y(x, y) = \frac{1}{2x + 3y}\left[-\frac{\partial}{\partial y}(2x + 3y) + \frac{\partial}{\partial x}(y - x)\right] = \frac{-4}{2x + 3y}$$

is not a function of y alone. We conclude that there is no straightforward IF and that another method has to be tried.

We note that the equation is homogeneous in x and y and so we set $y = vx$, with $\frac{\partial y}{\partial x} = v + x\frac{\partial v}{\partial x}$, and obtain

$$v + x\frac{\partial v}{\partial x} = -\frac{2 + 3v}{v - 1},$$

$$x\frac{\partial v}{\partial x} = \frac{-2 - 3v - v^2 + v}{v - 1} = -\frac{v^2 + 2v + 2}{v - 1},$$

$$\frac{dx}{x} = \frac{(1 - v)\,dv}{v^2 + 2v + 2}$$

$$= \frac{2}{(v + 1)^2 + 1} - \frac{v + 1}{(v + 1)^2 + 1},$$

$$\Rightarrow \quad \ln Ax = 2\tan^{-1}(v + 1) - \tfrac{1}{2}\ln[\,1 + (v + 1)^2\,],$$

$$\ln\left\{Bx^2[\,1 + (v + 1)^2\,]\right\} = 4\tan^{-1}(v + 1).$$

On setting $v = y/x$ this becomes

$$B[\,x^2 + (y + x)^2\,] = \exp\left[4\tan^{-1}\left(\frac{y + x}{x}\right)\right],$$

the final form of the solution.

14.13 Solve

$$\frac{dy}{dx} = -\frac{x+y}{3x+3y-4}.$$

Since x and y only appear in the combination $x+y$ we set $v = x+y$ with $dv/dx = 1 + dy/dx$. The equation and its solution then become

$$\frac{dv}{dx} = 1 - \frac{v}{3v-4},$$

$$dx = \frac{3v-4}{2v-4}dv = \left(\frac{3}{2} + \frac{2}{2v-4}\right)dv,$$

$$\Rightarrow \quad x+k = \tfrac{3}{2}v + \ln(v-2) = \tfrac{3}{2}(x+y) + \ln(x+y-2),$$

$$\ln(x+y-2) = k - \tfrac{1}{2}(x+3y).$$

Although the initial equation might look as if it could be made exact with an IF, applying the method described in Problem 14.3 shows that this not so; $B^{-1}[\partial A/\partial y - \partial B/\partial x]$ is neither zero nor a function of only one of the variables.

14.15 Find the curve with the property that at each point on it the sum of the intercepts on the x- and y-axes of the tangent to the curve (taking account of sign) is equal to unity.

At a point (X, Y) on the curve, the tangent to the curve is the straight line given by

$$y - Y = p(x - X),$$

where p is the slope of the tangent. This meets the axis $y = 0$ at $x = X - (Y/p)$ and the axis $x = 0$ at $y = Y - pX$. Thus, taking account of signs (i.e. some intercepts could be negative), the condition to be satisfied is

$$X - \frac{Y}{p} + Y - pX = 1.$$

Since (X, Y) lies on the required curve, the curve has an equation that satisfies

$$x - \frac{y}{p} + y - px = 1 \quad \Rightarrow \quad y = \frac{1-x+px}{1-p^{-1}} \quad (*).$$

Differentiating both sides of $(*)$ with respect to x, we now eliminate y by using the fact that its derivative with respect to x is p:

$$p = \frac{(1-p^{-1})(-1+p+xp') - (1-x+px)p^{-2}p'}{(1-p^{-1})^2},$$

$$p(p-1)^2 = (p^2-p)(p-1) + p'[x(p^2-p) - 1 + x - px].$$

The LHS and the first term on the RHS are equal, and so we have that either $p' = 0$ or

$$x(p^2 - 2p + 1) - 1 = 0,$$

$$\Rightarrow \quad x = \frac{1}{(p-1)^2},$$

$$\Rightarrow \quad p = 1 \pm \frac{1}{\sqrt{x}}.$$

From this and ($*$) it follows that

$$y = \frac{p[(1-x) + px]}{p-1} = \frac{\left(1 \pm \dfrac{1}{\sqrt{x}}\right)(1 - x + x \pm \sqrt{x})}{\pm \dfrac{1}{\sqrt{x}}}$$

$$= (\pm\sqrt{x} + 1)(1 \pm \sqrt{x}).$$

As expected, the solution is symmetric between x and y; this is demonstrated by the following rearrangement of the form just obtained:

$$y = (1 \pm \sqrt{x})^2,$$

$$\pm\sqrt{y} = 1 \pm \sqrt{x} \qquad (\pm \text{ signs not correlated}),$$

$$\pm\sqrt{y} - 1 = \pm\sqrt{x},$$

$$(1 \mp \sqrt{y})^2 = x.$$

Because of the square roots involved, a real curve exists only for x and y both positive, i.e. in the first quadrant. That curve is $\sqrt{x} + \sqrt{y} = 1$.

The singular solution $p' = 0$ (ignored earlier) corresponds to a set of curves, on each of which the slope is a constant. Any one such curve is a *straight* line joining the axial points $(\theta, 0)$ and $(0, 1 - \theta)$ for any arbitrary real θ; the tangent at any point on such a 'curve' is always the curve itself, whose intercepts, θ and $1 - \theta$, sum to unity.

14.17 Find the general solutions of the following:

(a) $\dfrac{dy}{dx} + \dfrac{xy}{a^2 + x^2} = x;$ (b) $\dfrac{dy}{dx} = \dfrac{4y^2}{x^2} - y^2.$

(a) With dy/dx appearing in the first term and y in the second (and nowhere else), this is a linear first-order ODE and therefore has an IF given by

$$\mu(x) = \exp\left\{\int \frac{x}{a^2 + x^2}\right\} dx = \exp[\tfrac{1}{2}\ln(a^2 + x^2)] = (a^2 + x^2)^{1/2}.$$

When multiplied through by this, the equation becomes

$$\frac{d}{dx}[(a^2 + x^2)^{1/2}y] = x(a^2 + x^2)^{1/2},$$

$$\Rightarrow \quad (a^2 + x^2)^{1/2}y = \tfrac{2}{3}\tfrac{1}{2}(a^2 + x^2)^{3/2} + A,$$

$$\Rightarrow \quad y = \frac{a^2 + x^2}{3} + \frac{A}{(a^2 + x^2)^{1/2}}.$$

(b) The RHS can be written as the product of one function of x and another one of y; the equation is therefore separable:

$$\frac{dy}{y^2} = \left(\frac{4}{x^2} - 1\right) dx,$$

$$\Rightarrow \quad -\frac{1}{y} = -\frac{4}{x} - x + A,$$

$$\Rightarrow \quad y = \frac{x}{x^2 + Bx + 4},$$

where $B = -A$ and is the arbitrary integration constant.

14.19 An electronic system has two inputs, to each of which a constant unit signal is applied, but starting at different times. The equations governing the system thus take the form

$$\dot{x} + 2y = H(t),$$

$$\dot{y} - 2x = H(t - 3).$$

Initially (at $t = 0$), $x = 1$ and $y = 0$; find $x(t)$ at later times.

Since we have coupled equations, working with their Laplace transforms suggests itself. This will convert the equations into simultaneous algebraic equations – though there may be some difficulty in converting the solution back into t-space.

The transform of the Heaviside function is s^{-1}, and so the two transformed equations (incorporating the initial conditions and using the translation property of Laplace transforms) are

$$s\bar{x} - 1 + 2\bar{y} = \frac{1}{s},$$

$$s\bar{y} - 0 - 2\bar{x} = \frac{1}{s}e^{-3s}.$$

Since it is $x(t)$ that we require, we eliminate $\bar{y}$ to obtain

$$s^2\bar{x} - s + \frac{2}{s}e^{-3s} + 4\bar{x} = 1,$$

from which

$$\bar{x} = \frac{s^2 + s - 2e^{-3s}}{s(s^2 + 4)},$$

$$= \frac{s+1}{s^2+4} + \left[-\frac{1}{2s} + \frac{s}{2(s^2+4)} \right] e^{-3s}.$$

For the first term in square brackets, the coefficient in the partial fractions expansion was determined by considering the limit $s \to 0$; that for the second term was found by inspection.

Now, using a look-up table if necessary, we find that, in t-space, the function corresponding to the $\bar{x}$ found above is

$$x(t) = \tfrac{1}{2} \sin 2t + \cos 2t - \tfrac{1}{2} H(t-3) + \tfrac{1}{2} H(t-3) \cos 2(t-3).$$

14.21 A reflecting mirror is made in the shape of the surface of revolution generated by revolving the curve $y(x)$ about the x-axis. In order that light rays emitted from a point source at the origin are reflected back parallel to the x-axis, the curve $y(x)$ must obey

$$\frac{y}{x} = \frac{2p}{1 - p^2},$$

where $p = dy/dx$. By solving this equation for x, find the curve $y(x)$.

The original equation can be rewritten as

$$y = \frac{2px}{1 - p^2}.$$

This can now be differentiated with respect to x to give (denoting dp/dx by p')

$$p = \frac{dy}{dx} = \frac{(1 - p^2)(2p + 2xp') - 2px(-2p)p'}{(1 - p^2)^2},$$

$$p(1 - p^2)^2 = 2p(1 - p^2) + 2x(1 - p^2)p' + 4p^2 x p',$$

$$p(1 - p^2)(-p^2 - 1) = 2x(1 + p^2)p',$$

$$-p(1 - p^2) = 2xp'.$$

The equation is now separable, and after expressing the p-dependent part in partial fractions we have

$$\frac{dx}{x} = \frac{2}{p(p^2 - 1)} = \frac{1}{p-1} + \frac{1}{p+1} - \frac{2}{p},$$

$$\ln x = \ln(p-1) + \ln(p+1) - 2\ln p + c,$$

$$x = A \frac{p^2 - 1}{p^2}.$$

From this we can write $p = \pm(1 - A^{-1}x)^{-1/2}$, but we first substitute for $1 - p^2$ in the original equation, obtaining

$$\frac{y}{x} = -\frac{2p}{A^{-1}p^2x} \quad \Rightarrow \quad y = -\frac{2A}{p} = \mp 2A(1 - A^{-1}x)^{1/2}.$$

Finally, squaring this result, we have

$$y^2 = 4A^2(1 - A^{-1}x) = 4A^2 - 4Ax;$$

this is the equation of a parabola with its apex at $x = A$.

14.23 Find the solution $y = y(x)$ of

$$x\frac{dy}{dx} + y - \frac{y^2}{x^{3/2}} = 0,$$

subject to $y(1) = 1$.

After being divided through by x, this equation is in the form of a Bernoulli equation with $n = 2$, i.e. it is of the form

$$\frac{dy}{dx} + P(x)y = Q(x)y^n.$$

Here, $P(x) = x^{-1}$ and $Q(x) = x^{-5/2}$. So we set $v = y^{1-2} = y^{-1}$ and obtain

$$\frac{dy}{dx} = \frac{d}{dx}\left(\frac{1}{v}\right) = -\frac{1}{v^2}\frac{dv}{dx}.$$

The equation then becomes

$$-\frac{1}{v^2}\frac{dv}{dx} + \frac{1}{vx} = \frac{1}{v^2 x^{5/2}},$$

$$\frac{dv}{dx} - \frac{v}{x} = -\frac{1}{x^{5/2}}, \quad \text{for which the IF is } 1/x,$$

$$\frac{d}{dx}\left(\frac{v}{x}\right) = -\frac{1}{x^{7/2}},$$

$$\frac{v}{x} = \frac{2}{5}\frac{1}{x^{5/2}} + \frac{3}{5}, \quad \text{using } y(1) = 1,$$

$$\frac{1}{y} = \frac{2}{5}\frac{1}{x^{3/2}} + \frac{3x}{5},$$

$$y = \frac{5x^{3/2}}{2 + 3x^{5/2}}.$$

14.25 Find the family of solutions of

$$\frac{d^2 y}{dx^2} + \left(\frac{dy}{dx}\right)^2 + \frac{dy}{dx} = 0$$

that satisfy $y(0) = 0$.

As the equation contains only derivatives, we write $dy/dx = p$ and $d^2 y/dx^2 = dp/dx$; this will reduce the equation to one of first order:

$$\frac{dp}{dx} + p^2 + p = 0.$$

Separating the variables:

$$\frac{dp}{p(p+1)} = -dx.$$

We now integrate and express the integrand in partial fractions:

$$\int \left(\frac{1}{p} - \frac{1}{p+1}\right) dp = -\int dx,$$

$$\ln(p) - \ln(p+1) = A - x,$$

$$\Rightarrow \quad \frac{p}{p+1} = Be^{-x},$$

$$\Rightarrow \quad p = \frac{e^{-x}}{C - e^{-x}}.$$

Now $p = dy/dx$ and so

$$\frac{dy}{dx} = \frac{e^{-x}}{C - e^{-x}},$$

$$y = \ln(C - e^{-x}) + D$$

$$= \ln(C - e^{-x}) - \ln(C - 1), \quad \text{since we require } y(0) = 0,$$

$$= \ln \frac{C - e^{-x}}{C - 1}.$$

This is as far as y can be determined, since only one boundary condition is given for a second-order equation. As C is varied, the solution generates a family of curves satisfying the original equation.

A variety of other forms of solution are possible and equally valid, the actual form obtained depending on where in the calculation the boundary condition is incorporated. They include

$$e^y = F(1 - e^{-x}) + 1, \quad y = \ln[\, G - (G - 1)e^{-x}\,], \quad y = \ln(e^{-K} + 1 - e^{-x}) + K.$$

14.27 A simple harmonic oscillator, of mass m and natural frequency ω_0, experiences an oscillating driving force $f(t) = ma \cos \omega t$. Therefore, its equation of motion is

$$\frac{d^2 x}{dt^2} + \omega_0^2 x = a \cos \omega t,$$

where x is its position. Given that at $t = 0$ we have $x = dx/dt = 0$, find the function $x(t)$. Describe the solution if ω is approximately, but not exactly, equal to ω_0.

To find the full solution given the initial conditions, we need the complete general solution made up of a complementary function (CF) and a particular integral (PI). The CF is clearly of the form $A \cos \omega_0 t + B \sin \omega_0 t$ and, in view of the form of the RHS, we try $x(t) = C \cos \omega t + D \sin \omega t$ as a PI. Substituting this gives

$$-\omega^2 C \cos \omega t - \omega^2 D \sin \omega t + \omega_0^2 C \cos \omega t + \omega_0^2 D \sin \omega t = a \cos \omega t.$$

Equating coefficients of the independent functions $\cos \omega t$ and $\sin \omega t$ requires that

$$-\omega^2 C + \omega_0^2 C = a \quad \Rightarrow \quad C = \frac{a}{\omega_0^2 - \omega^2},$$

$$-\omega^2 D + \omega_0^2 D = 0 \quad \Rightarrow \quad D = 0.$$

Thus, the general solution is

$$x(t) = A \cos \omega_0 t + B \sin \omega_0 t + \frac{a}{\omega_0^2 - \omega^2} \cos \omega t.$$

The initial conditions impose the requirements

$$x(0) = 0 \quad \Rightarrow \quad 0 = A + \frac{a}{\omega_0^2 - \omega^2},$$

$$\text{and } \dot{x}(0) = 0 \quad \Rightarrow \quad 0 = \omega_0 B.$$

Incorporating the implications of these into the general solution gives

$$x(t) = \frac{a}{\omega_0^2 - \omega^2} (\cos \omega t - \cos \omega_0 t)$$

$$= \frac{2a \sin[\frac{1}{2}(\omega + \omega_0)t] \sin[\frac{1}{2}(\omega_0 - \omega)t]}{(\omega_0 + \omega)(\omega_0 - \omega)}.$$

For $\omega_0 - \omega = \epsilon$ with $|\epsilon| t \ll 1$,

$$x(t) \approx \frac{2a \sin \omega_0 t \, \frac{1}{2} \epsilon t}{2\omega_0 \epsilon} = \frac{at}{2\omega_0} \sin \omega_0 t.$$

Thus, for moderate t, $x(t)$ is a sine wave of linearly increasing amplitude.

Over a long time, $x(t)$ will vary between $\pm 2a/(\omega_0^2 - \omega^2)$ with sizeable intervals between the two extremes, i.e. it will show beats of amplitude $2a/(\omega_0^2 - \omega^2)$.

14.29 The theory of bent beams shows that at any point in the beam the 'bending moment' is given by K/ρ, where K is a constant (that depends upon the beam material and cross-sectional shape) and ρ is the radius of curvature at that point. Consider a light beam of length L whose ends, $x = 0$ and $x = L$, are supported at the same vertical height and which has a weight W suspended from its centre. Verify that at any point x ($0 \le x \le L/2$ for definiteness) the net magnitude of the bending moment (bending moment = force × perpendicular distance) due to the weight and support reactions, evaluated on either side of x, is $Wx/2$.

If the beam is only slightly bent, so that $(dy/dx)^2 \ll 1$, where $y = y(x)$ is the downward displacement of the beam at x, show that the beam profile satisfies the approximate equation

$$\frac{d^2y}{dx^2} = -\frac{Wx}{2K}.$$

By integrating this equation twice and using physically imposed conditions on your solution at $x = 0$ and $x = L/2$, show that the downward displacement at the centre of the beam is $WL^3/(48K)$.

The upward reaction of the support at each end of the beam is $\frac{1}{2}W$. At the position x, with $0 \le x \le L/2$, the contributing moments are as follows.

The moment on the left is due to

(i) the support at $x = 0$ providing a clockwise moment of $\frac{1}{2}Wx$.

The moment on the right is due to

(ii) the support at $x = L$ providing an anticlockwise moment of $\frac{1}{2}W(L - x)$;
(iii) the weight at $x = \frac{1}{2}L$ providing a clockwise moment of $W(\frac{1}{2}L - x)$.

The net clockwise moment on the right is therefore $W(\frac{1}{2}L - x) - \frac{1}{2}W(L - x) = -\frac{1}{2}Wx$, i.e. equal in magnitude, but opposite in sign, to that on the left.

The radius of curvature of the beam is $\rho = [\,1 + (-y')^2\,]^{3/2}/(-y'')$, but if $|y'| \ll 1$ this simplifies to $-1/y''$ and the equation of the beam profile satisfies

$$\frac{Wx}{2} = M = \frac{K}{\rho} = -K\frac{d^2y}{dx^2}.$$

We now need to integrate this, taking into account the boundary conditions $y(0) = 0$ and, on symmetry grounds, $y'(\frac{1}{2}L) = 0$:

$$y' = -\frac{Wx^2}{4K} + A, \text{ with } y'(\tfrac{1}{2}L) = 0 \quad \Rightarrow \quad A = \frac{WL^2}{16K},$$

$$y' = \frac{W}{4K}\left(\frac{L^2}{4} - x^2\right),$$

$$y = \frac{W}{4K}\left(\frac{L^2 x}{4} - \frac{x^3}{3} + B\right), \text{ with } y(0) = 0 \quad \Rightarrow \quad B = 0.$$

The centre is lowered by

$$y(\tfrac{1}{2}L) = \frac{W}{4K}\left(\frac{L^2}{4}\frac{L}{2} - \frac{1}{3}\frac{L^3}{8}\right) = \frac{WL^3}{48K}.$$

Note that the derived analytic form for $y(x)$ is not applicable in the range $\frac{1}{2}L \leq x \leq L$; the beam profile is symmetrical about $x = \frac{1}{2}L$, but the expression $\frac{1}{4}L^2x - \frac{1}{3}x^3$ is not invariant under the substitution $x \to L - x$.

14.31 The function $f(t)$ satisfies the differential equation

$$\frac{d^2f}{dt^2} + 8\frac{df}{dt} + 12f = 12e^{-4t}.$$

For the following sets of boundary conditions determine whether it has solutions, and, if so, find them:

(a) $f(0) = 0,$ $f'(0) = 0,$ $f(\ln\sqrt{2}) = 0$;
(b) $f(0) = 0,$ $f'(0) = -2,$ $f(\ln\sqrt{2}) = 0$.

Three boundary conditions have been given, and, as this is a second-order linear equation for which only two independent conditions are needed, they may be inconsistent. The plan is to solve it using two of the conditions and then test whether the third one is compatible.
    The auxiliary equation for obtaining the CF is

$$m^2 + 8m + 12 = 0 \quad \Rightarrow \quad m = -2 \text{ or } m = -6$$

$$\Rightarrow \quad f(t) = Ae^{-6t} + Be^{-2t}.$$

Since the form of the RHS, Ce^{-4t}, is not included in the CF, we can try it as the particular integral:

$$16C - 32C + 12C = 12 \quad \Rightarrow \quad C = -3.$$

The general solution is therefore

$$f(t) = Ae^{-6t} + Be^{-2t} - 3e^{-4t}.$$

(a) For boundary conditions $f(0) = 0,$ $f'(0) = 0,$ $f(\ln\sqrt{2}) = 0$:

$$f(0) = 0 \quad \Rightarrow \quad A + B - 3 = 0,$$

$$f'(0) = 0 \quad \Rightarrow \quad -6A - 2B + 12 = 0,$$

$$\Rightarrow \quad A = \tfrac{3}{2}, \quad B = \tfrac{3}{2}.$$

$$\text{Hence, } f(t) \quad = \quad \tfrac{3}{2}e^{-6t} + \tfrac{3}{2}e^{-2t} - 3e^{-4t}.$$

Recalling that $e^{-(\ln\sqrt{2})} = 1/\sqrt{2}$, we evaluate

$$f(\ln\sqrt{2}) = \frac{3}{2}\frac{1}{8} + \frac{3}{2}\frac{1}{2} - 3\frac{1}{4} = \frac{3}{16} \neq 0.$$

Thus the boundary conditions are inconsistent and there is no solution.

(b) For boundary conditions $f(0) = 0$, $f'(0) = -2$, $f(\ln \sqrt{2}) = 0$, we proceed as before:

$$f(0) = 0 \quad \Rightarrow \quad A + B - 3 = 0,$$
$$f'(0) = 0 \quad \Rightarrow \quad -6A - 2B + 12 = -2,$$
$$\Rightarrow \quad A = 2, \quad B = 1.$$
$$\text{Hence, } f(t) \quad = \quad 2e^{-6t} + e^{-2t} - 3e^{-4t}.$$

We again evaluate

$$f(\ln \sqrt{2}) = 2\frac{1}{8} + \frac{1}{2} - 3\frac{1}{4} = 0.$$

This time the boundary conditions are consistent and there is a unique solution as given above.

14.33 A solution of the differential equation

$$\frac{d^2 y}{dx^2} + 2\frac{dy}{dx} + y = 4e^{-x}$$

takes the value 1 when $x = 0$ and the value e^{-1} when $x = 1$. What is its value when $x = 2$?

The auxiliary equation, $m^2 + 2m + 1 = 0$, has repeated roots $m = -1$, and so the general CF has the special form $y(x) = (A + Bx)e^{-x}$.

Turning to the PI, we note that the form of the RHS of the original equation is contained in the CF, and (to make matters worse) so is x times the RHS. We therefore need to take x^2 times the RHS as a trial PI:

$$y(x) = Cx^2 e^{-x}, \quad y' = C(2x - x^2)e^{-x}, \quad y'' = C(2 - 4x + x^2)e^{-x}.$$

Substituting these into the original equation shows that

$$2Ce^{-x} = 4e^{-x} \quad \Rightarrow \quad C = 2$$

and that the full general solution is given by

$$y(x) = (A + Bx)e^{-x} + 2x^2 e^{-x}.$$

We now determine the unknown constants using the information given about the solution. Since $y(0) = 1$, $A = 1$. Further, $y(1) = e^{-1}$ requires

$$e^{-1} = (1 + B)e^{-1} + 2e^{-1} \quad \Rightarrow \quad B = -2.$$

Finally, we conclude that $y(x) = (1 - 2x + 2x^2)e^{-x}$ and, therefore, that $y(2) = 5e^{-2}$.

14.35 Find the general solutions of

(a) $\dfrac{d^3y}{dx^3} - 12\dfrac{dy}{dx} + 16y = 32x - 8,$

(b) $\dfrac{d}{dx}\left(\dfrac{1}{y}\dfrac{dy}{dx}\right) + (2a \coth 2ax)\left(\dfrac{1}{y}\dfrac{dy}{dx}\right) = 2a^2$, where a is a constant.

(a) As this is a third-order equation, we expect three terms in the CF.

Since it is linear with constant coefficients, we can make use of the auxiliary equation, which is

$$m^3 - 12m + 16 = 0.$$

By inspection, $m = 2$ is one root; the other two can be found by factorisation:

$$m^3 - 12m + 16 = (m - 2)(m^2 + 2m - 8) = (m - 2)(m + 4)(m - 2) = 0.$$

Thus we have one repeated root ($m = 2$) and one other ($m = -4$) leading to a CF of the form

$$y(x) = (A + Bx)e^{2x} + Ce^{-4x}.$$

As the RHS contains no exponentials, we try $y(x) = Dx + E$ for the PI. We then need $16D = 32$ and $-12D + 16E = -8$, giving $D = 2$ and $E = 1$.

The general solution is therefore

$$y(x) = (A + Bx)e^{2x} + Ce^{-4x} + 2x + 1.$$

(b) The equation is already arranged in the form

$$\frac{dg(y)}{dx} + h(x)g(y) = j(x)$$

and so needs only an IF to allow the first integration step to be made. For this equation the IF is

$$\exp\left\{\int 2a \coth 2ax \, dx\right\} = \exp(\ln \sinh 2ax) = \sinh 2ax.$$

After multiplication through by this factor, the equation can be written

$$\sinh 2ax \frac{d}{dx}\left(\frac{1}{y}\frac{dy}{dx}\right) + (2a \cosh 2ax)\left(\frac{1}{y}\frac{dy}{dx}\right) = 2a^2 \sinh 2ax,$$

$$\frac{d}{dx}\left(\sinh 2ax \frac{1}{y}\frac{dy}{dx}\right) = 2a^2 \sinh 2ax.$$

Integrating this gives

$$\sinh 2ax \frac{1}{y}\frac{dy}{dx} = \frac{2a^2}{2a}\cosh 2ax + A,$$

$$\Rightarrow \quad \frac{1}{y}\frac{dy}{dx} = a \coth 2ax + \frac{A}{\sinh 2ax}.$$

Integrating again, $\quad \ln y = \dfrac{1}{2}\ln(\sinh 2ax) + \displaystyle\int \dfrac{A}{\sinh 2ax}\,dx + B$

$$= \dfrac{1}{2}\ln(\sinh 2ax) + \dfrac{A}{2a}\ln(|\tanh ax|) + B,$$

$$\Rightarrow \quad y = C(\sinh 2ax)^{1/2}\,(|\tanh ax|)^{D}.$$

The indefinite integral of $(\sinh 2ax)^{-1}$ appearing in the fourth line can be verified by differentiating $y = \ln|\tanh ax|$ in the form $y = \frac{1}{2}\ln(\tanh^{2} ax)$ and recalling that

$$\cosh ax \sinh ax = \dfrac{1}{2}\sinh 2ax.$$

14.37 The quantities $x(t)$, $y(t)$ satisfy the simultaneous equations

$$\ddot{x} + 2n\dot{x} + n^{2}x = 0,$$

$$\ddot{y} + 2n\dot{y} + n^{2}y = \mu\dot{x},$$

where $x(0) = y(0) = \dot{y}(0) = 0$ and $\dot{x}(0) = \lambda$. Show that

$$y(t) = \tfrac{1}{2}\mu\lambda t^{2}\left(1 - \tfrac{1}{3}nt\right)\exp(-nt).$$

For these two coupled equations, in which an 'output' from the first acts as the 'driving input' for the second, we take Laplace transforms and incorporate the boundary conditions:

$$(s^{2}\bar{x} - 0 - \lambda) + 2n(s\bar{x} - 0) + n^{2}\bar{x} = 0,$$

$$(s^{2}\bar{y} - 0 - 0) + 2n(s\bar{y} - 0) + n^{2}\bar{y} = \mu(s\bar{x} - 0).$$

From the first transformed equation,

$$\bar{x} = \dfrac{\lambda}{s^{2} + 2ns + n^{2}}.$$

Substituting this into the second transformed equation gives

$$\bar{y} = \dfrac{\mu s\bar{x}}{(s+n)^{2}} = \dfrac{\mu\lambda s}{(s+n)^{2}(s+n)^{2}}$$

$$= \dfrac{\mu\lambda}{(s+n)^{3}} - \dfrac{\mu\lambda n}{(s+n)^{4}},$$

$$\Rightarrow \quad y(t) = \mu\lambda\left(\dfrac{t^{2}}{2!}e^{-nt} - \dfrac{nt^{3}}{3!}e^{-nt}\right), \text{ from Table 13.1 in the main text,}$$

$$= \dfrac{1}{2}\mu\lambda t^{2}\left(1 - \dfrac{nt}{3}\right)e^{-nt}, \text{ i.e. as stated in the question.}$$

14.39 Two unstable isotopes A and B and a stable isotope C have the following decay rates per atom present: $A \to B$, $3\,\text{s}^{-1}$; $A \to C$, $1\,\text{s}^{-1}$; $B \to C$, $2\,\text{s}^{-1}$. Initially a quantity x_0 of A is present but there are no atoms of the other two types. Using Laplace transforms, find the amount of C present at a later time t.

Using the name symbol to represent the corresponding number of atoms and taking Laplace transforms, we have

$$\frac{dA}{dt} = -(3+1)A \quad \Rightarrow \quad s\bar{A} - x_0 = -4\bar{A}$$

$$\Rightarrow \quad \bar{A} = \frac{x_0}{s+4},$$

$$\frac{dB}{dt} = 3A - 2B \quad \Rightarrow \quad s\bar{B} = 3\bar{A} - 2\bar{B}$$

$$\Rightarrow \quad \bar{B} = \frac{3x_0}{(s+2)(s+4)},$$

$$\frac{dC}{dt} = A + 2B \quad \Rightarrow \quad s\bar{C} = \bar{A} + 2\bar{B}$$

$$\Rightarrow \quad \bar{C} = \frac{x_0(s+2) + 6x_0}{s(s+2)(s+4)}.$$

Using the 'cover-up' method for finding the coefficients of a partial fraction expansion without repeated factors, e.g. the coefficient of $(s+2)^{-1}$ is $[(-2+8)x_0]/[(-2)(-2+4)] = -6x_0/4$, we have

$$\bar{C} = \frac{x_0(s+8)}{s(s+2)(s+4)} = \frac{x_0}{s} - \frac{6x_0}{4(s+2)} + \frac{4x_0}{8(s+4)}$$

$$\Rightarrow \quad C(t) = x_0 \left(1 - \tfrac{3}{2}e^{-2t} + \tfrac{1}{2}e^{-4t}\right).$$

This is the required expression.

14.41 The 'golden mean', which is said to describe the most aesthetically pleasing proportions for the sides of a rectangle (e.g. the ideal picture frame), is given by the limiting value of the ratio of successive terms of the Fibonacci series u_n, which is generated by

$$u_{n+2} = u_{n+1} + u_n,$$

with $u_0 = 0$ and $u_1 = 1$. Find an expression for the general term of the series and verify that the golden mean is equal to the larger root of the recurrence relation's characteristic equation.

The recurrence relation is second order and its characteristic equation, obtained by setting $u_n = A\lambda^n$, is

$$\lambda^2 - \lambda - 1 = 0 \quad \Rightarrow \quad \lambda = \tfrac{1}{2}(1 \pm \sqrt{5}).$$

The general solution is therefore

$$u_n = A\left(\frac{1+\sqrt{5}}{2}\right)^n + B\left(\frac{1-\sqrt{5}}{2}\right)^n.$$

The initial values (boundary conditions) determine A and B:

$$u_0 = 0 \quad \Rightarrow \quad B = -A,$$

$$u_1 = 1 \quad \Rightarrow \quad A\left(\frac{1+\sqrt{5}}{2} - \frac{1-\sqrt{5}}{2}\right) = 1 \quad \Rightarrow \quad A = \frac{1}{\sqrt{5}},$$

$$\text{Hence, } u_n \quad = \quad \frac{1}{\sqrt{5}}\left[\left(\frac{1+\sqrt{5}}{2}\right)^n - \left(\frac{1-\sqrt{5}}{2}\right)^n\right].$$

If we write $(1 - \sqrt{5})/(1 + \sqrt{5}) = r < 1$, the ratio of successive terms in the series is

$$\frac{u_{n+1}}{u_n} \quad = \quad \frac{\frac{1}{2}[(1+\sqrt{5})^{n+1} - (1-\sqrt{5})^{n+1}]}{(1+\sqrt{5})^n - (1-\sqrt{5})^n}$$

$$= \quad \frac{\frac{1}{2}[1+\sqrt{5} - (1-\sqrt{5})r^n]}{1 - r^n}$$

$$\rightarrow \quad \frac{1+\sqrt{5}}{2} \text{ as } n \rightarrow \infty;$$

i.e. the limiting ratio is the same as the larger value of λ.

This result is a particular example of the more general one that the ratio of successive terms in a series generated by a recurrence relation tends to the largest (in absolute magnitude) of the roots of the characteristic equation. Here there are only two roots, but for an Nth-order relation there will be N roots.

14.43 The first few terms of a series u_n, starting with u_0, are $1, 2, 2, 1, 6, -3$. The series is generated by a recurrence relation of the form

$$u_n = Pu_{n-2} + Qu_{n-4},$$

where P and Q are constants. Find an expression for the general term of the series and show that, in fact, the series consists of two interleaved series given by

$$u_{2m} = \tfrac{2}{3} + \tfrac{1}{3}4^m,$$

$$u_{2m+1} = \tfrac{7}{3} - \tfrac{1}{3}4^m,$$

for $m = 0, 1, 2, \ldots$

We first find P and Q using

$$n = 4 \qquad 6 = 2P + Q,$$

$$n = 5 \quad -3 = P + 2Q, \quad \Rightarrow \quad Q = -4 \text{ and } P = 5.$$

The recurrence relation is thus

$$u_n = 5u_{n-2} - 4u_{n-4}.$$

To solve this we try $u_n = A + B\lambda^n$ for arbitrary constants A and B and obtain

$$A + B\lambda^n = 5A + 5B\lambda^{n-2} - 4A - 4B\lambda^{n-4},$$

$$\Rightarrow \quad 0 = \lambda^4 - 5\lambda^2 + 4$$

$$= (\lambda^2 - 1)(\lambda^2 - 4) \quad \Rightarrow \quad \lambda = \pm 1, \pm 2.$$

The general solution is $\quad u_n = A + B(-1)^n + C2^n + D(-2)^n.$

We now need to solve the simultaneous equations for A, B, C and D provided by the values of $u_0, \ldots, u_3$:

$$1 = A + B + C + D,$$

$$2 = A - B + 2C - 2D,$$

$$2 = A + B + 4C + 4D,$$

$$1 = A - B + 8C - 8D.$$

These have the straightforward solution

$$A = \frac{3}{2}, \quad B = -\frac{5}{6}, \quad C = \frac{1}{12}, \quad D = \frac{1}{4},$$

and so

$$u_n = \frac{3}{2} - \frac{5}{6}(-1)^n + \frac{1}{12}2^n + \frac{1}{4}(-2)^n.$$

When n is even and equal to $2m$,

$$u_{2m} = \frac{3}{2} - \frac{5}{6} + \frac{4^m}{12} + \frac{4^m}{4} = \frac{2}{3} + \frac{4^m}{3}.$$

When n is odd and equal to $2m + 1$,

$$u_{2m+1} = \frac{3}{2} + \frac{5}{6} + \frac{4^m}{6} - \frac{4^m}{2} = \frac{7}{3} - \frac{4^m}{3}.$$

In passing, we note that the fact that both P and Q, and all of the given values $u_0, \ldots, u_4$, are integers, and hence that all terms in the series are integers, provides an indirect proof that $4^m + 2$ is divisible by 3 (without remainder) for all non-negative integers m. This can be more easily proved by induction, as the reader may like to verify.

14.45 Find the general expression for the u_n satisfying

$$u_{n+1} = 2u_{n-2} - u_n$$

with $u_0 = u_1 = 0$ and $u_2 = 1$, and show that they can be written in the form

$$u_n = \frac{1}{5} - \frac{2^{n/2}}{\sqrt{5}} \cos\left(\frac{3\pi n}{4} - \phi\right),$$

where $\tan \phi = 2$.

The characteristic equation (which will be a cubic since the recurrence relation is third order) and its solution are given by

$$\lambda^{n+1} = 2\lambda^{n-2} - \lambda^n,$$

$$\lambda^3 + \lambda^2 - 2 = 0,$$

$$(\lambda - 1)(\lambda^2 + 2\lambda + 2) = 0 \quad \Rightarrow \quad \lambda = 1 \text{ or } \lambda = -1 \pm i.$$

Thus the general solution of the recurrence relation, which has the generic form $A\lambda_1^n + B\lambda_2^n + C\lambda_3^n$, is

$$u_n = A + B(-1 + i)^n + C(-1 - i)^n$$

$$= A + B\, 2^{n/2} e^{i3\pi n/4} + C\, 2^{n/2} e^{i5\pi n/4}.$$

To determine A, B and C we use

$$u_0 = 0, \qquad 0 = A + B + C,$$

$$u_1 = 0, \qquad 0 = A + B\, 2^{1/2} e^{i3\pi/4} + C\, 2^{1/2} e^{i5\pi/4}$$

$$= A + B(-1 + i) + C(-1 - i),$$

$$u_2 = 1, \qquad 1 = A + B\, 2e^{i6\pi/4} + C\, 2e^{i10\pi/4} = A + 2B(-i) + 2C(i).$$

Adding twice each of the first two equations to the last one gives $5A = 1$. Substituting this into the first and last equations then leads to

$$B + C = -\frac{1}{5} \quad \text{and} \quad -B + C = \frac{2}{5i},$$

from which it follows that

$$B = \frac{-1 + 2i}{10} = \frac{\sqrt{5}}{10} e^{i(\pi - \phi)}$$

$$\text{and} \quad C = \frac{-1 - 2i}{10} = \frac{\sqrt{5}}{10} e^{i(\pi + \phi)},$$

where $\tan \phi = 2/1 = 2$.

Thus, collecting these results together, we have

$$u_n = \frac{1}{5} + \frac{2^{n/2}\sqrt{5}}{10} [e^{i3\pi n/4} e^{i(\pi - \phi)} + e^{i5\pi n/4} e^{i(\pi + \phi)}]$$

$$= \frac{1}{5} - \frac{2^{n/2}\sqrt{5}}{10} (e^{i3\pi n/4} e^{-i\phi} + e^{-i3\pi n/4} e^{i\phi})$$

$$= \frac{1}{5} - \frac{2^{n/2}\sqrt{5}}{10} \left[2\cos\left(\frac{3\pi n}{4} - \phi\right) \right]$$

$$= \frac{1}{5} - \frac{2^{n/2}}{\sqrt{5}} \cos\left(\frac{3\pi n}{4} - \phi\right),$$

i.e. the form of solution given in the question.

15 Elementary probability

15.1 By shading or numbering Venn diagrams, determine which of the following are valid relationships between events. For those that are, prove the relationship using de Morgan's laws.

(a) $\overline{(\bar{X} \cup Y)} = X \cap \bar{Y}$.
(b) $\bar{X} \cup \bar{Y} = \overline{(X \cup Y)}$.
(c) $(X \cup Y) \cap Z = (X \cup Z) \cap Y$.
(d) $X \cup \overline{(Y \cap Z)} = (X \cup \bar{Y}) \cap \bar{Z}$.
(e) $X \cup \overline{(Y \cap Z)} = (X \cup \bar{Y}) \cup \bar{Z}$.

For each part of this question we refer to the corresponding part of Figure 15.1.

(a) This relationship is correct as both expressions define the shaded region that is both inside X and outside Y.

(b) This relationship is *not* valid. The LHS specifies the whole sample space *apart from* the region marked with the heavy shading. The RHS defines the region that is lightly shaded. The unmarked regions of X and Y are included in the former but not in the latter.

(c) This relationship is *not* valid. The LHS specifies the sum of the regions marked 2, 3 and 4 in the figure, whilst the RHS defines the sum of the regions marked 1, 3 and 4.

(d) This relationship is *not* valid. On the LHS, $\overline{Y \cap Z}$ is the whole sample space apart from regions 3 and 4. So $X \cup \overline{(Y \cap Z)}$ consists of all regions except for region 3. On the RHS, $X \cup \bar{Y}$ contains all regions except 3 and 7. The events $\bar{Z}$ contain regions 1, 6, 7 and 8 and so $(X \cup \bar{Y}) \cap \bar{Z}$ consists of regions 1, 6 and 8. Thus regions 2, 4, 5 and 7 are in one specification but not in the other.

(e) This relationship is valid. The LHS is as found in (d), namely all regions except for region 3. The RHS consists of the union (as opposed to the intersection) of the two subregions found in (d) and thus contains those regions found in either or both of $X \cup \bar{Y}$ (1, 2, 4, 5, 6 and 8) and $\bar{Z}$ (1, 6, 7 and 8). This covers all regions except region 3 – in agreement with those found for the LHS.

For the two valid relationships, their proofs using de Morgan's laws and the associativity of the union operator are:

(a) $\overline{(\bar{X} \cup Y)} = \bar{\bar{X}} \cap \bar{Y} = X \cap \bar{Y}$,
(e) $X \cup \overline{(Y \cap Z)} = X \cup (\bar{Y} \cup \bar{Z}) = (X \cup \bar{Y}) \cup \bar{Z}$.

Here we have also used the result that the complement of the complement of a set is the set itself.

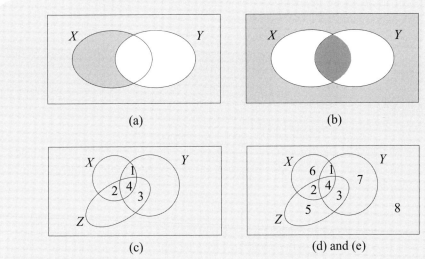

(a)

(b)

(c)

(d) and (e)

Figure 15.1 The Venn diagrams used in Problem 15.1

15.3   *A* and *B* each have two unbiased four-faced dice, the four faces being numbered 1, 2, 3 and 4. Without looking, *B* tries to guess the sum *x* of the numbers on the bottom faces of *A*'s two dice after they have been thrown onto a table. If the guess is correct *B* receives x^2 euros, but if not he loses *x* euros. Determine *B*'s expected gain per throw of *A*'s dice when he adopts each of the following strategies:

(a) He selects *x* at random in the range $2 \le x \le 8$.
(b) He throws his own two dice and guesses *x* to be whatever they indicate.
(c) He takes your advice and always chooses the same value for *x*. Which number would you advise?

We first calculate the probabilities $p(x)$ and the corresponding gains $g(x) = p(x)x^2 - [1 - p(x)]x$ for each value of the total *x*. Expressing both in units of $1/16$, they are as follows:

x	2	3	4	5	6	7	8
$p(x)$	1	2	3	4	3	2	1
$g(x)$	-26	-24	-4	40	30	0	-56

(a) If *B*'s guess is random in the range $2 \le x \le 8$ then his expected return is

$$\frac{1}{16}\frac{1}{7}(-26 - 24 - 4 + 40 + 30 + 0 - 56) = -\frac{40}{112} = -0.36 \text{ euros.}$$

(b) If he picks by throwing his own dice then his distribution of guesses is the same as that of $p(x)$ and his expected return is

$$\frac{1}{16}\frac{1}{16}[1(-26) + 2(-24) + 3(-4) + 4(40) + 3(30) + 2(0) + 1(-56)]$$

$$= \frac{108}{256} = 0.42 \text{ euros.}$$

(c) If B chooses y, then his expected return is $h(y) = p(y)y^2 - \sum_{x \neq y} p(x)x$. An additional line in the table (in the same units) would read $h(x)$, -74, -56, -20, 40, 46, 32, -8. You should not advise B, but take his place, guess '6' each time, and expect an average profit of $46/16$ euros.

15.5 Two duellists, A and B, take alternate shots at each other, and the duel is over when a shot (fatal or otherwise!) hits its target. Each shot fired by A has a probability α of hitting B, and each shot fired by B has a probability β of hitting A. Calculate the probabilities P_1 and P_2, defined as follows, that A will win such a duel: P_1, A fires the first shot; P_2, B fires the first shot.
    If they agree to fire simultaneously, rather than alternately, what is the probability P_3 that A will win, i.e. hit B without being hit himself?

Each shot has only two possible outcomes, a hit or a miss. P_1 is the probability that A will win when it is his turn to fire the next shot, and he is still able to do so (event W). There are three possible outcomes of the first two shots: C_1, A hits with his shot; C_2, A misses but B hits; C_3, both miss. Thus

$$P_1 = \sum_i \Pr(C_i) \Pr(W|C_i)$$

$$= [\alpha \times 1] + [(1 - \alpha)\beta \times 0] + [(1 - \alpha)(1 - \beta) \times P_1]$$

$$\Rightarrow \quad P_1 = \frac{\alpha}{\alpha + \beta - \alpha\beta}.$$

When B fires first but misses, the situation is the one just considered. But if B hits with his first shot then clearly A's chances of winning are zero. Since these are the only two possible outcomes of B's first shot, we can write

$$P_2 = [\beta \times 0] + [(1 - \beta) \times P_1] \quad \Rightarrow \quad P_2 = \frac{(1 - \beta)\alpha}{\alpha + \beta - \alpha\beta}.$$

When both fire at the same time there are four possible outcomes D_i to the first round: D_1, A hits and B misses; D_2, B hits but A misses; D_3, they both hit; D_4, they both miss. If getting hit, even if you manage to hit your opponent, does not count as a win, then

$$P_3 = \sum_i \Pr(D_i) \Pr(W|D_i)$$

$$= [\alpha(1 - \beta) \times 1] + [(1 - \alpha)\beta \times 0] + [\alpha\beta \times 0] + [(1 - \alpha)(1 - \beta) \times P_3].$$

This can be rearranged as

$$P_3 = \frac{\alpha(1 - \beta)}{\alpha + \beta - \alpha\beta} = P_2.$$

Thus the result is the same as if B had fired first. However, we also note that if all that matters to A is that B is hit, whether or not he is hit himself, then the third bracket takes the value $\alpha\beta \times 1$ and P_3 takes the same value as P_1.

15.7 An electronics assembly firm buys its microchips from three different suppliers; half of them are bought from firm X, whilst firms Y and Z supply 30% and 20%, respectively. The suppliers use different quality-control procedures and the percentages of defective chips are 2%, 4% and 4% for X, Y and Z, respectively. The probabilities that a defective chip will fail two or more assembly-line tests are 40%, 60% and 80%, respectively, whilst all defective chips have a 10% chance of escaping detection. An assembler finds a chip that fails only one test. What is the probability that it came from supplier X?

Since the number of tests failed by a defective chip are mutually exclusive outcomes (0, 1 or ≥ 2), a chip supplied by X has a probability of failing just one test given by $0.02(1 - 0.1 - 0.4) = 0.010$. The corresponding probabilities for chips supplied by Y and Z are $0.04(1 - 0.1 - 0.6) = 0.012$ and $0.04(1 - 0.1 - 0.8) = 0.004$, respectively.

Using '1' to denote failing a single test, Bayes' theorem gives the probability that the chip was supplied by X as

$$\Pr(X|1) = \frac{\Pr(1|X)\,\Pr(X)}{\Pr(1|X)\,\Pr(X) + \Pr(1|Y)\,\Pr(Y) + \Pr(1|Z)\,\Pr(Z)}$$

$$= \frac{0.010 \times 0.5}{0.010 \times 0.5 + 0.012 \times 0.3 + 0.004 \times 0.2} = \frac{50}{94}.$$

15.9 A boy is selected at random from amongst the children belonging to families with n children. It is known that he has at least two sisters. Show that the probability that he has $k - 1$ brothers is

$$\frac{(n - 1)!}{(2^{n-1} - n)(k - 1)!(n - k)!},$$

for $1 \leq k \leq n - 2$ and zero for other values of k. Assume that boys and girls are equally likely.

The boy has $n - 1$ siblings. Let A_j be the event that $j - 1$ of them are brothers, i.e. his family contains j boys and $n - j$ girls. The probability of event A_j is

$$\Pr(A_j) = \frac{^{n-1}C_{j-1}\left(\frac{1}{2}\right)^{n-1}}{\sum_{j=1}^{n}\,^{n-1}C_{j-1}\left(\frac{1}{2}\right)^{n-1}} = \frac{(n - 1)!}{2^{n-1}(j - 1)!(n - j)!}.$$

If B is the event that the boy has at least two sisters, then

$$\Pr(B|A_j) = \begin{cases} 1 & 1 \leq j \leq n - 2, \\ 0 & n - 1 \leq j \leq n. \end{cases}$$

Now we apply Bayes' theorem to give the probability that he has $k - 1$ brothers:

$$\Pr(A_k|B) = \frac{1\ \Pr(A_k)}{\sum_{j=1}^{n-2} 1\ \Pr(A_j)},$$

for $1 \le k \le n - 2$. The denominator of this expression is the sum $1 = (\frac{1}{2} + \frac{1}{2})^{n-1} = \sum_{j=1}^{n}\ ^{n-1}C_{j-1}\left(\frac{1}{2}\right)^{n-1}$, but omitting the $j = n - 1$ and the $j = n$ terms, and so is equal to

$$1 - \frac{(n-1)!}{2^{n-1}(n-2)!\,1!} - \frac{(n-1)!}{2^{n-1}(n-1)!\,0!} = \frac{1}{2^{n-1}}\left[2^{n-1} - (n-1) - 1\right].$$

Thus,

$$\Pr(A_k|B) = \frac{(n-1)!}{2^{n-1}(k-1)!(n-k)!}\frac{2^{n-1}}{2^{n-1}-n} = \frac{(n-1)!}{(2^{n-1}-n)(k-1)!(n-k)!},$$

as given in the question.

15.11 A set of $2N + 1$ rods consists of one of each integer length $1, 2, \ldots, 2N, 2N + 1$. Three, of lengths a, b and c, are selected, of which a is the longest. By considering the possible values of b and c, determine the number of ways in which a non-degenerate triangle (i.e. one of non-zero area) can be formed (i) if a is even and (ii) if a is odd. Combine these results appropriately to determine the total number of non-degenerate triangles that can be formed using three of the $2N + 1$ rods, and hence show that the probability that such a triangle can be formed from a random selection (without replacement) of three rods is

$$\frac{(N-1)(4N+1)}{2(4N^2-1)}.$$

Rod a is the longest of the three rods. As no two are the same length, let $a > b > c$. To form a non-degenerate triangle we require that $b + c > a$, and, in consequence, $4 \le a \le 2N + 1$.

(i) With a even. Consider each b ($< a$) in turn and determine how many values of c allow a triangle to be made:

b	Values of c	Number of c values
$a - 1$	$2, 3, \cdots, a - 2$	$a - 3$
$a - 2$	$3, 4, \cdots, a - 3$	$a - 5$
$\cdots$	$\cdots$	$\cdots$
$\frac{1}{2}a + 1$	$\frac{1}{2}a$	1

Thus there are $1 + 3 + 5 + \cdots + (a - 3)$ possible triangles when a is even.

(ii) A table for odd a is similar, except that the last line will read $b = \frac{1}{2}(a + 3)$, $c = \frac{1}{2}(a - 1)$ or $\frac{1}{2}(a + 1)$, and the number of c values $= 2$. Thus there are $2 + 4 + 6 + \cdots + (a - 3)$ possible triangles when a is odd.

To find the total number $n(N)$ of possible triangles, we group together the cases $a = 2m$ and $a = 2m + 1$, where $m = 1, 2, \ldots, N$. Then,

$$n(N) = \sum_{m=2}^{N} [\, 1 + 3 + \cdots + (2m - 3)\,] + [\, 2 + 4 + \cdots + (2m + 1 - 3)\,]$$

$$= \sum_{m=2}^{N} \sum_{k=1}^{2m-2} k = \sum_{m=2}^{N} \tfrac{1}{2}(2m - 2)(2m - 1) = \sum_{m=2}^{N} 2m^2 - 3m + 1$$

$$= 2\left[\tfrac{1}{6}N(N + 1)(2N + 1) - 1\right] - 3\left[\tfrac{1}{2}N(N + 1) - 1\right] + N - 1$$

$$= \frac{N}{6}[\, 2(N + 1)(2N + 1) - 9(N + 1) + 6\,]$$

$$= \frac{N}{6}(4N^2 - 3N - 1) = \frac{N}{6}(4N + 1)(N - 1).$$

The number of ways that three rods can be drawn at random (without replacement) is $(2N + 1)(2N)(2N - 1)/3!$ and so the probability that they can form a triangle is

$$\frac{N(4N + 1)(N - 1)}{6} \frac{3!}{(2N + 1)(2N)(2N - 1)} = \frac{(N - 1)(4N + 1)}{2(4N^2 - 1)},$$

as stated in the question.

15.13 The duration (in minutes) of a telephone call made from a public call-box is a random variable T. The probability density function (PDF) of T is

$$f(t) = \begin{cases} 0 & t < 0, \\ \frac{1}{2} & 0 \leq t < 1, \\ ke^{-2t} & t \geq 1, \end{cases}$$

where k is a constant. To pay for the call, 20 pence has to be inserted at the beginning, and a further 20 pence after each subsequent half-minute. Determine by how much the average cost of a call exceeds the cost of a call of average length charged at 40 pence per minute.

From the normalisation of the PDF, we must have

$$1 = \int_0^\infty f(t)\, dt = \frac{1}{2} + \int_1^\infty ke^{-2t}\, dt = \frac{1}{2} + \frac{ke^{-2}}{2} \quad \Rightarrow \quad k = e^2.$$

The average length of a call is given by

$$\bar{t} = \int_0^1 t\,\frac{1}{2}\, dt + \int_1^\infty t\, e^2 e^{-2t}\, dt$$

$$= \frac{1}{2}\frac{1}{2} + \left[\frac{te^2 e^{-2t}}{-2}\right]_1^\infty + \int_1^\infty \frac{e^2 e^{-2t}}{2}\, dt = \frac{1}{4} + \frac{1}{2} + \frac{e^2}{2}\left[\frac{e^{-2t}}{-2}\right]_1^\infty = \frac{3}{4} + \frac{1}{4} = 1.$$

Let $p_n = \Pr\{\frac{1}{2}(n - 1) < t < \frac{1}{2}n\}$. The corresponding cost is $c_n = 20n$.

Clearly, $p_1 = p_2 = \frac{1}{4}$ and, for $n > 2$,

$$p_n = e^2 \int_{(n-1)/2}^{n/2} e^{-2t}\, dt = e^2 \left[\frac{e^{-2t}}{-2} \right]_{(n-1)/2}^{n/2} = \frac{1}{2}e^2(e-1)e^{-n}.$$

The average cost of a call is therefore

$$\bar{c} = 20\left[\frac{1}{4} + 2\frac{1}{4} + \sum_{n=3}^{\infty} n\frac{1}{2}e^2(e-1)e^{-n} \right] = 15 + 10e^2(e-1)\sum_{n=3}^{\infty} ne^{-n}.$$

Now, the final summation might be recognised as part of an arithmetico-geometric series whose sum can be found from the standard formula

$$S = \frac{a}{1-r} + \frac{rd}{(1-r)^2},$$

with $a = 0, d = 1$ and $r = e^{-1}$, or could be evaluated directly by noting that as a geometric series,

$$\sum_{n=0}^{\infty} e^{-nx} = \frac{1}{1-e^{-x}}.$$

Differentiating this with respect to x and then setting $x = 1$ gives

$$-\sum_{n=0}^{\infty} ne^{-nx} = -\frac{e^{-x}}{(1-e^{-x})^2} \quad \Rightarrow \quad \sum_{n=0}^{\infty} ne^{-n} = \frac{e^{-1}}{(1-e^{-1})^2}.$$

From either method it follows that

$$\sum_{n=3}^{\infty} ne^{-n} = \frac{e}{(e-1)^2} - e^{-1} - 2e^{-2}$$

$$= \frac{e - e + 2 - e^{-1} - 2 + 4e^{-1} - 2e^{-2}}{(e-1)^2} = \frac{3e^{-1} - 2e^{-2}}{(e-1)^2}.$$

The total charge therefore exceeds that of a call of average length (1 minute) charged at 40 pence per minute by the amount (in pence)

$$15 + 10e^2(e-1)\frac{3e^{-1} - 2e^{-2}}{(e-1)^2} - 40 = \frac{10(3e-2) - 25e + 25}{e-1} = \frac{5e+5}{e-1} = 10.82.$$

15.15 A tennis tournament is arranged on a straight knockout basis for 2^n players, and for each round, except the final, opponents for those still in the competition are drawn at random. The quality of the field is so even that in any match it is equally likely that either player will win. Two of the players have surnames that begin with 'Q'. Find the probabilities that they play each other

(a) in the final,
(b) at some stage in the tournament.

Let p_r be the probability that *before* the rth round the two players are both still in the tournament (and, by implication, have not met each other). Clearly, $p_1 = 1$.

Before the rth round there are 2^{n+1-r} players left in. For both 'Q' players to still be in before the $(r + 1)$th round, Q_1 must avoid Q_2 in the draw and both must win their matches. Thus

$$p_{r+1} = \frac{2^{n+1-r} - 2}{2^{n+1-r} - 1} \left(\frac{1}{2}\right)^2 p_r.$$

(a) The probability that they meet in the final is p_n, given by

$$p_n = 1 \frac{2^n - 2}{2^n - 1} \frac{1}{4} \frac{2^{n-1} - 2}{2^{n-1} - 1} \frac{1}{4} \cdots \frac{2^2 - 2}{2^2 - 1} \frac{1}{4}$$

$$= \left(\frac{1}{4}\right)^{n-1} 2^{n-1} \left[\frac{(2^{n-1} - 1)(2^{n-2} - 1) \cdots (2^1 - 1)}{(2^n - 1)(2^{n-1} - 1) \cdots (2^2 - 1)}\right]$$

$$= \left(\frac{1}{4}\right)^{n-1} 2^{n-1} \frac{1}{2^n - 1}$$

$$= \frac{1}{2^{n-1}(2^n - 1)}.$$

(b) The more general solution to the recurrence relation derived above is

$$p_r = 1 \frac{2^n - 2}{2^n - 1} \frac{1}{4} \frac{2^{n-1} - 2}{2^{n-1} - 1} \frac{1}{4} \cdots \frac{2^{n+2-r} - 2}{2^{n+2-r} - 1} \frac{1}{4}$$

$$= \left(\frac{1}{4}\right)^{r-1} 2^{r-1} \left[\frac{(2^{n-1} - 1)(2^{n-2} - 1) \cdots (2^{n+1-r} - 1)}{(2^n - 1)(2^{n-1} - 1) \cdots (2^{n+2-r} - 1)}\right]$$

$$= \left(\frac{1}{2}\right)^{r-1} \frac{2^{n+1-r} - 1}{2^n - 1}.$$

Before the rth round, if they are both still in the tournament, the probability that they will be drawn against each other is $(2^{n-r+1} - 1)^{-1}$. Consequently, the chance that they will meet at *some* stage is

$$\sum_{r=1}^{n} p_r \frac{1}{2^{n-r+1} - 1} = \sum_{r=1}^{n} \left(\frac{1}{2}\right)^{r-1} \frac{2^{n+1-r} - 1}{2^n - 1} \frac{1}{2^{n-r+1} - 1}$$

$$= \frac{1}{2^n - 1} \sum_{r=1}^{n} \left(\frac{1}{2}\right)^{r-1}$$

$$= \frac{1}{2^n - 1} \frac{1 - (\frac{1}{2})^n}{1 - \frac{1}{2}} = \frac{1}{2^{n-1}}.$$

This same conclusion can also be reached in the following way. The probability that Q_1 *is not* put out of (i.e. wins) the tournament is $(\frac{1}{2})^n$. It follows that the probability that Q_1 *is* put out is $1 - (\frac{1}{2})^n$ and that the player responsible is Q_2 with probability $[1 - (\frac{1}{2})^n]/(2^n - 1) = 2^{-n}$. Similarly, the probability that Q_2 is put out and that the player responsible is Q_1 is also 2^{-n}. These are exclusive events but cover all cases in which Q_1 and Q_2 meet during the tournament, the probability of which is therefore $2 \times 2^{-n} = 2^{n-1}$.

15.17 A point P is chosen at random on the circle $x^2 + y^2 = 1$. The random variable X denotes the distance of P from $(1, 0)$. Find the mean and variance of X and the probability that X is greater than its mean.

With O as the centre of the unit circle and Q as the point $(1, 0)$, let OP make an angle θ with the x-axis OQ. The random variable X then has the value $2 \sin(\theta/2)$ with θ uniformly distributed on $(0, 2\pi)$, i.e.

$$f(x)\,dx = \frac{1}{2\pi}\,d\theta.$$

The mean of X is given straightforwardly by

$$\langle X \rangle = \int_0^2 X f(x)\,dx = \int_0^{2\pi} 2\sin\left(\frac{\theta}{2}\right)\frac{1}{2\pi}\,d\theta = \frac{1}{\pi}\left[-2\cos\frac{\theta}{2}\right]_0^{2\pi} = \frac{4}{\pi}.$$

For the variance we have

$$\sigma_X^2 = \langle X^2 \rangle - \langle X \rangle^2 = \int_0^{2\pi} 4\sin^2\left(\frac{\theta}{2}\right)\frac{1}{2\pi}\,d\theta - \frac{16}{\pi^2} = \frac{4}{2\pi}\frac{1}{2}2\pi - \frac{16}{\pi^2} = 2 - \frac{16}{\pi^2}.$$

When $X = \langle X \rangle = 4/\pi$, the angle $\theta = 2\sin^{-1}(2/\pi)$ and so

$$\Pr(X > \langle X \rangle) = \frac{2\pi - 4\sin^{-1}\dfrac{2}{\pi}}{2\pi} = 0.561.$$

15.19 The number of errors needing correction on each page of a set of proofs follows a Poisson distribution of mean μ. The cost of the first correction on any page is α and that of each subsequent correction on the same page is β. Prove that the average cost of correcting a page is

$$\alpha + \beta(\mu - 1) - (\alpha - \beta)e^{-\mu}.$$

Since the number of errors on a page is Poisson distributed, the probability of n errors on any particular page is

$$\Pr(n \text{ errors}) = p_n = e^{-\mu}\frac{\mu^n}{n!}.$$

The average cost per page, found by averaging the corresponding cost over all values of n, is

$$c = 0\,p_0 + \alpha p_1 + \sum_{n=2}^{\infty} [\alpha + (n-1)\beta]\,p_n$$

$$= \alpha\mu e^{-\mu} + (\alpha - \beta)\sum_{n=2}^{\infty} p_n + \beta \sum_{n=2}^{\infty} np_n.$$

Now, $\sum_{n=0}^{\infty} p_n = 1$ and, for a Poisson distribution, $\sum_{n=0}^{\infty} np_n = \mu$. These can be used to evaluate the above, once the $n = 0$ and $n = 1$ terms have been removed. Thus

$$c = \alpha\mu e^{-\mu} + (\alpha - \beta)(1 - e^{-\mu} - \mu e^{-\mu}) + \beta(\mu - 0 - \mu e^{-\mu})$$

$$= \alpha + \beta(\mu - 1) + e^{-\mu}(\alpha\mu - \alpha + \beta - \mu\alpha + \mu\beta - \mu\beta)$$

$$= \alpha + \beta(\mu - 1) + e^{-\mu}(\beta - \alpha),$$

as given in the question.

15.21 The probability distribution for the number of eggs in a clutch is Po(λ), and the probability that each egg will hatch is p (independently of the size of the clutch). Show by direct calculation that the probability distribution for the number of chicks that hatch is Po(λp).

Clearly, to determine the probability that a clutch produces k chicks, we must consider clutches of size n, for all $n \geq k$, and for each such clutch find the probability that exactly k of the n chicks do hatch. We then average over all n, weighting the results according to the distribution of n.

The probability that k chicks hatch from a clutch of size n is $^nC_k p^k q^{n-k}$, where $q = 1 - p$. The probability that the clutch is of size n is $e^{-\lambda}\lambda^n/n!$. Consequently, the overall probability of k chicks hatching from a clutch is

$$\Pr(k\text{ chicks}) = \sum_{n=k}^{\infty} e^{-\lambda} \frac{\lambda^n}{n!}\ ^nC_k\,p^k\,q^{n-k}$$

$$= e^{-\lambda} p^k \lambda^k \sum_{n=k}^{\infty} \frac{(\lambda q)^{n-k}}{n!} \frac{n!}{k!\,(n-k)!}, \qquad \text{set } n - k = m,$$

$$= e^{-\lambda} \frac{(\lambda p)^k}{k!} \sum_{m=0}^{\infty} \frac{(\lambda q)^m}{m!}$$

$$= e^{-\lambda} \frac{(\lambda p)^k}{k!} e^{\lambda q}$$

$$= \frac{e^{-\lambda p}(\lambda p)^k}{k!},$$

since $q = 1 - p$. Thus $\Pr(k \text{ chicks})$ is distributed as a Poisson distribution with parameter $\mu = \lambda p$.

15.23 Under EU legislation on harmonisation, all kippers are to weigh 0.2000 kg and vendors who sell underweight kippers must be fined by their government. The weight of a kipper is normally distributed with a mean of 0.2000 kg and a standard deviation of 0.0100 kg. They are packed in cartons of 100 and large quantities of them are sold.

Every day a carton is to be selected at random from each vendor and tested according to one of the following schemes, which have been approved for the purpose.

(a) The entire carton is weighed and the vendor is fined 2500 euros if the average weight of a kipper is less than 0.1975 kg.
(b) Twenty-five kippers are selected at random from the carton; the vendor is fined 100 euros if the average weight of a kipper is less than 0.1980 kg.
(c) Kippers are removed one at a time, at random, until one has been found that weighs *more* than 0.2000 kg; the vendor is fined $4n(n - 1)$ euros, where n is the number of kippers removed.

Which scheme should the Chancellor of the Exchequer be urging his government to adopt?

For these calculations we measure weights in grams.

(a) For this scheme we have a normal distribution with mean $\mu = 200$ and s.d. $\sigma = 10$. The s.d. for a carton is $\sqrt{100}\,\sigma = 100$ and the mean weight is 20 000. There is a penalty if the weight of a carton is less than 19 750. This critical value represents a standard variable of

$$Z = \frac{19\,750 - 20\,000}{100} = -2.5.$$

The probability that $Z < -2.5 = 1 - \Phi(2.5) = 1 - 0.9938 = 0.0062$. Thus the average fine per carton tested on this scheme is $0.0062 \times 2500 = 15.5$ euros.

(b) For this scheme the general parameters are the same but the mean weight of the sample measured is 5000 and its s.d is $\sqrt{25}\,(10) = 50$. The Z-value at which a fine is imposed is

$$Z = \frac{(198 \times 25) - 5000}{50} = -1.$$

The probability that $Z < -1.0 = 1 - \Phi(1.0) = 1 - 0.8413 = 0.1587$. Thus the average fine per carton tested on this scheme is $0.1587 \times 100 = 15.9$ euros.

(c) This scheme is a series of Bernoulli trials in which the probability of success is $\frac{1}{2}$ (since half of all kippers weigh more than 200 and the distribution is normal). The probability that it will take n kippers to find one that passes the test is $q^{n-1}p = (\frac{1}{2})^n$. The expected fine is therefore

$$f = \sum_{n=2}^{\infty} 4n(n-1) \left(\frac{1}{2}\right)^n = 4\frac{2\left(\frac{1}{4}\right)}{\left(\frac{1}{2}\right)^3} = 16 \text{ euros.}$$

The expression for the sum was found by twice differentiating the sum of the geometric series $\sum r^n$ with respect to r, as follows:

$$\sum_{n=0}^{\infty} r^n = \frac{1}{1-r} \quad \Rightarrow \quad \sum_{n=1}^{\infty} n r^{n-1} = \frac{1}{(1-r)^2}$$

$$\Rightarrow \quad \sum_{n=2}^{\infty} n(n-1) r^{n-2} = \frac{2}{(1-r)^3}$$

$$\Rightarrow \quad \sum_{n=2}^{\infty} n(n-1) r^n = \frac{2r^2}{(1-r)^3}.$$

There is, in fact, little to choose between the schemes on monetary grounds; no doubt political considerations, such as the current unemployment rate, will decide!

15.25 A practical-class demonstrator sends his 12 students to the storeroom to collect apparatus for an experiment, but forgets to tell each which type of component to bring. There are three types, A, B and C, held in the stores (in large numbers) in the proportions 20%, 30% and 50%, respectively, and each student picks a component at random. In order to set up one experiment, one unit each of A and B and two units of C are needed. Let $\Pr(N)$ be the probability that at least N experiments can be setup.

(a) Evaluate $\Pr(3)$.
(b) Find an expression for $\Pr(N)$ in terms of k_1 and k_2, the numbers of components of types A and B, respectively, selected by the students. Show that $\Pr(2)$ can be written in the form

$$\Pr(2) = (0.5)^{12} \sum_{i=2}^{6} {}^{12}C_i \, (0.4)^i \sum_{j=2}^{8-i} {}^{12-i}C_j \, (0.6)^j.$$

(c) By considering the conditions under which no experiments can be set up, show that $\Pr(1) = 0.9145$.

(a) To make three experiments possible the 12 components picked must be three each of A and B and six of C. The probability of this is given by the multinomial distribution as

$$\Pr(3) = \frac{(12)!}{3!\,3!\,6!} (0.2)^3 (0.3)^3 (0.5)^6 = 0.06237.$$

(b) Let the numbers of A, B and C selected be k_1, k_2 and k_3, respectively, and consider when *at least* N experiments can be set up. We have the obvious inequalities $k_1 \geq N$, $k_2 \geq N$ and $k_3 \geq 2N$. In addition, $k_3 = 12 - k_1 - k_2$, implying that $k_2 \leq 12 - 2N - k_1$. Further, k_1 cannot be greater than $12 - 3N$ if at least N experiments are to be set up, as each requires three other components that are not of type A. These inequalities set the limits on the acceptable values of k_1 and k_2 (k_3 is not a third independent variable). Thus $\Pr(N)$ is given by

$$\sum_{k_1 \geq N}^{12-3N} \sum_{k_2 \geq N}^{12-2N-k_1} \frac{(12)!}{k_1!\,k_2!\,(12 - k_1 - k_2)!} (0.2)^{k_1} (0.3)^{k_2} (0.5)^{12-k_1-k_2}.$$

The answer to part (a) is a particular case of this with $N = 3$, when each summation reduces to a single term.

For $N = 2$ the expression becomes

$$\Pr(2) = \sum_{k_1 \geq 2}^{6} \sum_{k_2 \geq 2}^{8-k_1} \frac{(12)!}{k_1! \, k_2! \, (12 - k_1 - k_2)!} (0.2)^{k_1} (0.3)^{k_2} (0.5)^{12-k_1-k_2}$$

$$= (0.5)^{12} \sum_{i=2}^{6} \sum_{j=2}^{8-i} \frac{(12)! \, (0.2/0.5)^i}{i! \, (12-i)!} \frac{(12-i)! \, (0.3/0.5)^j}{j! \, (12-i-j)!}$$

$$= (0.5)^{12} \sum_{i=2}^{6} {}^{12}C_i \, (0.4)^i \sum_{j=2}^{8-i} {}^{12-i}C_j \, (0.6)^j.$$

(c) No experiment can be set up if any one of the following four events occurs: $A_1 = (k_1 = 0)$, $A_2 = (k_2 = 0)$, $A_3 = (k_3 = 0)$ and $A_4 = (k_3 = 1)$. The probability for the union of these four events is given by

$$\Pr(A_1 \cup A_2 \cup A_3 \cup A_4) = \sum_{i=1}^{4} \Pr(A_i) - \sum_{i,j} \Pr(A_i \cap A_j) + \cdots$$

The probabilities $\Pr(A_i)$ are straightforward to calculate as follows:

$$\Pr(A_1) = (1 - 0.2)^{12}, \qquad \Pr(A_2) = (1 - 0.3)^{12},$$
$$\Pr(A_3) = (1 - 0.5)^{12}, \qquad \Pr(A_4) = {}^{12}C_1(1 - 0.5)^{12}(0.5).$$

The calculation of the probability for the intersection of two events is typified by

$$\Pr(A_1 \cap A_2) = [\, 1 - (0.2 + 0.3) \,]^{12}$$
$$\text{and } \Pr(A_1 \cap A_4) = {}^{12}C_1[\, 1 - (0.2 + 0.5) \,]^{11}(0.5)^1.$$

A few trial evaluations show that these are of order 10^{-4} and can be ignored by comparison with the larger terms in the first sum, which are (after rounding)

$$\sum_{i=1}^{4} \Pr(A_i) = (0.8)^{12} + (0.7)^{12} + (0.5)^{12} + 12(0.5)^{11}(0.5)$$

$$= 0.0687 + 0.0138 + 0.0002 + 0.0029 = 0.0856.$$

Since the probability of no experiments being possible is 0.0856, it follows that $\Pr(1) = 0.9144$.

15.27 The continuous random variables X and Y have a joint PDF proportional to $xy(x - y)^2$ with $0 \leq x \leq 1$ and $0 \leq y \leq 1$. Find the marginal distributions for X and Y and show that they are negatively correlated with correlation coefficient $-\frac{2}{3}$.

This PDF is clearly symmetric between x and y. We start by finding its normalisation constant c:

$$\int_0^1 \int_0^1 c(x^3 y - 2x^2 y^2 + xy^3)\, dx\, dy = c\left(\frac{1}{4}\frac{1}{2} - 2\frac{1}{3}\frac{1}{3} + \frac{1}{2}\frac{1}{4}\right) = \frac{c}{36}.$$

Thus, we must have that $c = 36$.

The marginal distribution for x is given by

$$f(x) = 36\int_0^1 (x^3 y - 2x^2 y^2 + xy^3)\, dy$$
$$= 36(\tfrac{1}{2}x^3 - \tfrac{2}{3}x^2 + \tfrac{1}{4}x)$$
$$= 18x^3 - 24x^2 + 9x,$$

and the mean of x by

$$\mu_X = \bar{x} = \int_0^1 (18x^4 - 24x^3 + 9x^2)\, dx = \frac{18}{5} - \frac{24}{4} + \frac{9}{3} = \frac{3}{5}.$$

By symmetry, the marginal distribution and the mean for y are $18y^3 - 24y^2 + 9y$ and $\frac{3}{5}$, respectively.

To calculate the correlation coefficient we also need the variances of x and y and their covariance. The variances, obviously equal, are given by

$$\sigma_X^2 = \int_0^1 x^2(18x^3 - 24x^2 + 9x)\, dx - (\tfrac{3}{5})^2$$
$$= \frac{18}{6} - \frac{24}{5} + \frac{9}{4} - \frac{9}{25}$$
$$= \frac{900 - 1440 + 675 - 108}{300} = \frac{9}{100}.$$

The standard deviations σ_X and σ_Y are therefore both equal to $3/10$.

The covariance is calculated next; it is given by

$$\text{Cov}\,[X, Y] = \langle XY \rangle - \mu_X \mu_Y$$
$$= 36\int_0^1 \int_0^1 (x^4 y^2 - 2x^3 y^3 + x^2 y^4)\, dx\, dy - \frac{3}{5}\frac{3}{5}$$
$$= \frac{36}{5 \times 3} - \frac{72}{4 \times 4} + \frac{36}{3 \times 5} - \frac{9}{25}$$
$$= \frac{12}{5} - \frac{9}{2} + \frac{12}{5} - \frac{9}{25}$$
$$= \frac{120 - 225 + 120 - 18}{50} = -\frac{3}{50}.$$

Finally,

$$\text{Corr}\,[X, Y] = \frac{\text{Cov}\,[X, Y]}{\sigma_X \sigma_Y} = \frac{-\frac{3}{50}}{\frac{3}{10}\frac{3}{10}} = -\frac{2}{3}.$$

15.29 Two continuous random variables X and Y have a joint probability distribution

$$f(x, y) = A(x^2 + y^2),$$

where A is a constant and $0 \le x \le a, 0 \le y \le a$. Show that X and Y are negatively correlated with correlation coefficient $-15/73$. By sketching a rough contour map of $f(x, y)$ and marking off the regions of positive and negative correlation, convince yourself that this (perhaps counter-intuitive) result is plausible.

The calculations of the various parameters of the distribution are straightforward (see Problem 15.27). The parameter A is determined by the normalisation condition:

$$1 = \int_0^a \int_0^a A(x^2 + y^2) \, dx \, dy = A \left(\frac{a^4}{3} + \frac{a^4}{3} \right) \quad \Rightarrow \quad A = \frac{3}{2a^4}.$$

The two expectation values required are given by

$$E[X] = \int_0^a \int_0^a Ax(x^2 + y^2) \, dx \, dy$$

$$= \frac{3}{2a^4} \left(\frac{a^5}{4 \times 1} + \frac{a^5}{2 \times 3} \right) = \frac{5a}{8}, \qquad (E[Y] = E[X]),$$

$$E[X^2] = \int_0^a \int_0^a Ax^2(x^2 + y^2) \, dx \, dy$$

$$= \frac{3}{2a^4} \left(\frac{a^6}{5 \times 1} + \frac{a^6}{3 \times 3} \right) = \frac{7a^2}{15}.$$

Hence the variance, calculated from the general result $V[X] = E[X^2] - (E[X])^2$, is

$$V[X] = \frac{7a^2}{15} - \left(\frac{5a}{8} \right)^2 = \frac{73}{960} a^2,$$

and the standard deviations are given by

$$\sigma_X = \sigma_Y = \sqrt{\frac{73}{960}} \, a.$$

To obtain the correlation coefficient we need also to calculate the following:

$$E[XY] = \int_0^a \int_0^a Axy(x^2 + y^2) \, dx \, dy$$

$$= \frac{3}{2a^4} \left(\frac{a^6}{4 \times 2} + \frac{a^6}{2 \times 4} \right) = \frac{3a^2}{8}.$$

Then the covariance, given by Cov $[X, Y] = E[XY] - E[X]E[Y]$, is evaluated as

$$\text{Cov } [X, Y] = \frac{3}{8}a^2 - \frac{5a}{8} \frac{5a}{8} = -\frac{a^2}{64}.$$

Combining this last result with the standard deviations calculated above, we then obtain

$$\text{Corr}\,[X, Y] = \frac{-(a^2/64)}{\sqrt{\frac{73}{960}}\,a\,\sqrt{\frac{73}{960}}\,a} = -\frac{15}{73}.$$

As the means of both X and Y are $\frac{5}{8}a = 0.62a$, the areas of the square of side a for which $X - \mu_X$ and $Y - \mu_Y$ have the same sign (i.e. regions of positive correlation) are about $(0.62)^2 \approx 39\%$ and $(0.38)^2 \approx 14\%$ of the total area of the square. The regions of negative correlation occupy some 47% of the square.

However, $f(x, y) = A(x^2 + y^2)$ favours the regions where one or both of x and y are large and close to unity. Broadly speaking, this gives little weight to the region in which both X and Y are less than their means, and so, although it is the largest region in area, it contributes relatively little to the overall correlation. The two (equal area) regions of negative correlation together outweigh the smaller high probability region of positive correlation in the top right-hand corner of the square; the overall result is a net negative correlation coefficient.

A Physical constants

Speed of light in a vacuum, $c = 3.0 \times 10^8 \, \text{m s}^{-1}$.

Elementary charge, $e = 1.60 \times 10^{-19} \, \text{C}$.

Mass of electron, $m_e = 9.1 \times 10^{-31} \, \text{kg}$.

Mass of proton, $m_p = 1.67 \times 10^{-27} \, \text{kg}$.

Avogadro constant, $N_A = 6.0 \times 10^{23} \, \text{mol}^{-1}$.

Planck constant, $h = 6.6 \times 10^{-34} \, \text{J s}$.

Boltzmann constant, $k = 1.38 \times 10^{-23} \, \text{J K}^{-1}$.

Stefan–Boltzmann constant, $\sigma = 5.7 \times 10^{-8} \, \text{W m}^{-2} \, \text{K}^{-4}$.

Gravitational constant, $G = 6.7 \times 10^{-11} \, \text{N kg}^{-2} \, \text{m}^2$.

Gravitational acceleration $g \approx 9.8 \, \text{m s}^{-2}$.

Permeability of a vacuum, $\mu_0 = 4\pi \times 10^{-7} \, \text{H m}^{-1}$.

Permittivity of a vacuum, $\epsilon_0 = 8.8 \times 10^{-12} \, \text{F m}^{-1}$.